U35 Stories:

Alarm Stations in U35

Eng.-Lt. Hans Fechter

&

U35 on the Hunt

Lothar von Arnauld de la Perière

The Great War at Sea Series

Volumes in the Great War at Sea Series

V188 My Torpedo Boat War Missions. Friedrich von Callisen
The Submarine War 1914-1918. Andreas Michelsen
The Last Cruise of the Majestic. George Goodrich
Two Lone Ships. Georg Kopp
U-boats Westward! Ernst Hashagen
Six Years of Submarine Cruising. Johannes Spiess
German Submarines and their Warfare 1914-1918. Albert Gayer
U21 Saves the Dardanelles. Otto Hersing
By U-boat Through the World's Oceans. Paul Schultz
The Logbook of U66. Excerpts from the First Officer's letters and diary. Joachim Kramsta. Compiled by Elsa Kramsta
U-Cruiser 151. Karl Plath
Submarine Minelayers in the War. Paul Kagelmann
Werewolf of the Seas. Robert Moraht
U35 Stories: Alarm Stations in U35, Hans Fechter & U35 on the Hunt, Lothar von Arnauld de la Perière

Cover art: Hans Rudi Erdt's poster publicizing the BUFA (Bild- und Filmamt) U-boat propaganda film U-Boote heraus! Mit U-Boot 178 gegen den Feind (1917).

U35 Stories:

Alarm Stations in U35

Eng.-Lt. Hans Fechter

&

U35 on the Hunt

Lothar von Arnauld de la Perière

Translated by Celestino Corraliza

Published by Trident Publishing
Miami, Fl USA

Library of Congress Cataloging-in-Publication Data

Originally published as In der Alarmkoje von U35. Berlin. 1918.
U35 Stories: Alarm Stations in U35 & U35 on the Hunt
ISBN-13: 978-1-959764-64-9
World War, 1914-1918 -- Personal narratives, German.
World War, 1914-1918 -- Naval operations.
Military operations, Naval--Submarine.

Printed in the United States of America

Translator's Forward

War literature in the era of total warfare is generally viewed with caution as it is subject to censorship and the objective is to create propaganda and form public opinion. A number of publishers in wartime Germany produced titles intended to reveal aspects of the war by actual participants. Ullstein Verlag created a series of Kriegsbücher, War Books, of which this title formed one of 42 volumes published.[1]

One can verify these works' contents as there is now access to archives, official reports, postwar memoirs and other eyewitness accounts. In three wartime submarine memoir titles examined[2]-the events and technical information presented are accurate. Sentiments expressed are common and expected of relatively youthful Regular Navy officers. There is less posturing than contained in a number German postwar submarine memoirs grappling with defeat, career destruction, and perceived betrayal. The censor's hand is evident in cloaking operational intelligence which would be useful to adversaries.

Hans Fechter's narrative, In Der Alarmkoje von U35[3], is of interest on a variety of levels. U35 is the most successful submarine of the conflict, sinking a considerable amount of tonnage and shipping (226 vessels) which in itself creates interest in its operations. Fechter's Engineering Section

---

[1] The most famous title of the series is Der Rote Kampfflieger by Manfred von Richtofen which has been dissected extensively by First World War specialists since publication and found to be useful.

[2] Ullstein's Hans Fechter, In der Alarmkoje von U35; Walter Forstmann, U39 Auf Jagd in Mittlemeer and publisher A. Scherl's wartime Heino von Heimburg's U-boot gegen U-Boot.

[3] In U35's Alarm Bunk. The more understandable Alarm Stations is utilized here.

duty station also provides information into the functions of that specialization and the care of early diesels and related submarine equipment. Mechanical and technical efficiency was a critical component in a submarine's performance which translated into success in sinking ships and survivability.[4] Fechter was considered extremely capable in this regard, hence his selection to create a work which informs and shapes views.

The format of these wartime books is a focus on action, brevity and atmosphere so the work is highly selective of incidents in view of the extensive available combats to choose from.

There is one other standalone participant account of U35's Mediterranean activities.[5] Famed long term commander during these operations, Kapitänleutnant Lothar von Arnauld de la Perière, penned a brief work postwar which refers to a number of the chapters subjects in Fechter's work. His work is appended after the main text and thus two views are provided, one by a deck officer and one by an engineer officer. Note the amount of information disclosed in Fechter's wartime work when the same incidents are relayed in von Arnauld's postwar recollection. As per custom, footnotes are added where Entente/Neutral reports contribute information.

Hans Fechter and half of U35's crew followed von Arnauld to a new command, U139, and the work concludes with this new boat.

C.Corraliza

---

4 Based on the careers of commanders and submarines the other attributes would be weapons' performance, commander determination and Luck.

5 Von Arnauld was also interviewed in Lowell Thomas' Raiders of the Deep anthology which tends to narrate events in the era's journalism style.

Contents

Through Gibraltar to the Mediterranean

Holiday days have slipped by, full of sun, freedom and happiness - now the train rolls through the green, wide plains of northern Germany towards the sea and the harbor where my boat was moored in the shipyard for some work. The after-sound of parting slowly fades away as the train rolls; thoughts wander ahead — and at noon I feel the old planks of my boat under my feet again.

The work has progressed vigorously, now it's time to test all machine systems. Hundreds of questions, just as many orders; and in between checking everything that has been worked on. After two hours I'm back right in the middle of things as if I had never been outside.

The starboard engine, my pride, is about to turn for the first time when an orderly approaches me.

"Commander's[6] orders: Meeting with the department director at four o'clock. Herr Oberingenieur is to also be present!"

"Yes, good!" — "So don't turn the starboard side! Until I come back, continue working on the port engine!"

Out of the work clothes, quickly dressed up; it is already ten minutes to four o'clock. I ask myself: what does the department director want from you? It must be something important! But let's wait and see.

---

[6] Kapitänleutnant Waldemar Kophamel is the commanding office of U35 at this time.

Everything is already assembled in the meeting room—the stragglers arrive with me. Shortly thereafter, the director opens the meeting.

"Gentlemen, first of all I would like to know: by when can boats U35 and U34 be ready to leave?"

Shipyard and commanders agree on five days.

"Good. We have orders to equip the boats for their next voyage entirely according to the instructions given by the command. Do you gentlemen have any special requests?"

Has something like this already been experienced in the Imperial Navy? The boats can equip themselves as they wish; - everything is done as you want it - no authorities, no official channels, in short and to the point: what do you want?

Splendid!

Of course, requests flood in, everything is sought out that one would otherwise hardly dare to request in one's mind. Spare parts and modifications are requested - and the amazing thing is: everything is approved without hesitation.

But why? — All of a sudden this question pops up and immediately becomes the main point. There is something behind that willingness. We used to have to ask a long time about every little thing and write lengthy requests before anything was approved, and now all of a sudden everything we command is as good as done.

All the same, the current largess is the main thing for the time being and the opportunity must be seized. Unfortunately, it only lasts five short days, and this period sets a limit to overly lavish desires. However, after an hour long lists are in the hands of the builders; Workers come on board — as many as you want. But the question "Why?" and "What For?" is asked quietly inside. Nobody knows anything, everyone thinks, guesses, creates scenarios; nobody knows anything specific.

The five days rush by: the commander and chief engineer must have found out more by now, their orders are too specific and precise. The most unusual pieces of equipment are brought on board, non-commissioned officers and crews wonder and ask whether this or that thing is actually supposed to be with us. "Yes, it should be on board."

Of course, the wildest rumors spread — supplies are growing enormously, and after speculations and questions comes the realization: It will be a long journey!

Finally the hour of departure has come.

It's Sunday. Bright, laughing sunshine lies over the harbor. "Hohenzollern weather!" as we say in the Navy, and today His Majesty is in town to hand over our newest and largest battle cruiser to the watery elements. The cheers resound thunderously across the harbor, heralding the happy passage of the proud ship that has received Hindenburg's name.[7]

Then, far out, the mighty lock gates open, and our little gray boat slips unnoticed out into the North Sea, and steers toward that red rock, under whose guns we now lie, to England's fury, as calm as in peace.

Heligoland is our next destination, and here we practice and rehearse again: day and night the work does not rest. It is actually wrong to work continuously before a long-distance journey; but what's the point, we've been granted three days, and then we should be ready to go.

---

[7] The battlecruiser Hindenburg was launched on August 1, 1915 from Wihelmshaven naval dockyard.

Diving takes place, the oil machines are running and have to give their utmost because all the damage, down to the smallest, has to be repaired; better to break down in port than out at sea a thousand miles from home shore. But anything that breaks or is discovered to be defective during the day must be repaired or removed at night. However, we succeed - we manage - and on the fourth day everything is clear, the supplies are replenished, new things are brought on board again, and in the last hour the helmsman is sent away to fetch the sea charts.

Then it's ready for sea at nine o'clock. The commander comes on board at five minutes to nine. "Both engines are ready to sail — below deck is ready for diving," reports the chief engineer; "Upper deck is ready for sea and diving!" the officer of the watch; "Bridge and rudder ready!" the helmsman.

A short whistle: "Ready to maneuver!"

One long, one short: "Cast off!" The boat slowly pushes itself out to the harbor, followed shortly afterwards by U34, which is supposed to sail with us at the same time.

As soon as we are in open water, the commander's call echoes across deck: "Everyone inside!" and into the tower: "On diving stations, diving tanks on!"

Quickly, one after the other, they climb down the steep ladder and disappear inside, where they hurry to their diving station to open the diving tanks.

WHEET, the speaking tubes in the control room whistle —

"Attention!"

"Forward dive tanks are up!'"

WHEET! "Attention!"

"Aft diving tanks are open!" reports the engineer on duty from the aft compartment.

"Tower!"

"Attention!"

"Report to the commander: Diving tanks are open!" I call out through the sound tube transmitting to the tower. The helmsman repeats it and calls it to the commander, who is just climbing into the tower.

"Depth Control, Herr Obersteuermann!"

"Go to ten meters!"

The bow leans slightly lower and the boat steers evenly into the depths.

"Ten meters!" reports the helmsman upwards.

"Trim the boat!"a voice replies. "Forward periscope cannot be raised!"

"Insert a new fuse!"

"One hundred liters aft!" A moment later the trim pump hums for a moment.

"A hundred liters are aft!"

"Ready trim tanks!"

"Take in two hundred liters!"

"Bottom vents open!"

"Fifty liters" — "Hundred" — "One hundred and fifty" — "Two hundred!"

"Set!"

Two hundred liters are in.

"Both rudders amidships!" A look at the water level and the depth gauge: "Boat is trimmed!"

"Surface!"

The rudders make the boat stern-heavy and it quickly rises to the surface.

"Turn on the fan, blow it out!"

A whirring and humming rises - "Starter is on!" reports the Central watch non-commissioned officer.

"All vents open except for six and seven!" I shout over the noise. All the vents and valves must be operated just as precisely as a column of men marching must coordinate on the parade ground.

"It's being blown out!" Screeching, the blower pulls in the air and presses the water out of the tanks with it.

"Eleven and twelve, thirteen" - "Open vents six and seven!"

"Dive tank nine, four blows!" and shortly after:

"Everything!"

Again hands fly from valve to valve, a wave, the big blower switch is pulled out, the blower supply air ducts are closed and what has happened is reported.

"Fan is switched off! Diving tanks are secure!"

"Ready the starboard oil engine! Ahead standard!"

Less than a minute after the fan was switched off, the oil engines was already rattling and panting in the next room, and the screws slowly began to turn.

The test dive is over, the boat was found to be in order, now the journey can begin.

U34 also dives and checks if it is ready. After about ten minutes it comes back up and both boats start their journey together.

On the evening of this first day we stand by...

Here we run zigzag courses to avoid any enemy submarines and provide no possibility of attack.

U34 announces by semaphore his engine coupling is not quite in order. The lower the sun sinks, the hazier and thicker the weather becomes, until finally we are enveloped in dense fog.

This is a good thing, because an enemy vessel can only spot us at a short distance.

Enemy destroyers are not yet to be expected here; so we drive through.

This time the North Sea means well for us: the next day we have calm, beautiful weather and after a short test dive we continue our journey.

The sea around us remains desolate and empty as in "Tristan".[8] Now and then the smoke of a steamer, the masts of a sailing ship come into view. We avoid them all, because we have strict orders to reach our destination, the Mediterranean, in the shortest possible time. Since yesterday we all know it and every last man is proud of our new mission. To achieve this, we also have to resist our desire to attack and avoid everything.

My engines are running well. During the course of the day a seal flies out of a valve on one side; the engines are stopped and the damage is repaired. The cooling water

[8] Reference to Richard Wagner's opera Tristan und Isolde.

losses are still too great, but this deficiency can only be remedied gradually. For the time being, we have started the journey and have the determination to bring it to a successful conclusion, despite all the difficulties that may arise.

Nothing disturbs our journey the next night either, and in the morning we stand just under the coast of England and want to pass through the narrow passage. At ten o'clock in the morning both oil engines are therefore running "Full Ahead". We are enveloped in dense gusts of rain and only occasionally see cloudy, blurred land shimmering through.

Around noon the weather clears up; at one thirty the watch then spots the first destroyer on the horizon. He hasn't seen us yet, so we turn and try to get past him without diving. But it didn't take long before he noticed us and headed towards us. So we dived.

An hour later, nothing more can be seen of him through the periscope. We then surface and now discover not only the destroyer, but also two trawlers, which he has probably brought in to support him. There is nothing left but to dive again. They must have seen us during our brief stay on the surface, because they cling to us and won't let go. Only at seven o'clock do we think we can dare again and show up.

But our luck does not last long this time either, because after three quarters of an hour a destroyer is on our heels and pushes us under water again.

Finally, at eleven o'clock, with increasing darkness, we manage to get free from him. Of course, the electric battery, which feeds the motors during the underwater trip, has to be charged first, because we were under for around ten hours. At the same time, we want to get out of this uncomfortable corner as quickly as possible.

At two o'clock in the morning the battery is full and we can now free ourselves even more quickly from the uncomfortable tentacles.

However, as soon as dawn breaks, another small enemy cruiser is in sight, from which we have to dive at six o'clock.

First, the commander wants to attack him. But he steers in zigzag courses, so it's difficult to get to him, and the commander also discovers that he has a heavy search device on the stern. He tries to get away from him in the shortest possible way.

This happened at eleven o'clock in the morning. The English still have one more "Dido" class cruiser, and we too can now continue our way undisturbed. The Port oil engine has to drive and the other charge so that we can move forward and the battery remain full. An island is abeam.

I use the afternoon hours to spend a good few hours from two to four o'clock, because the previous night there had been little sleep, three single hours, that was the whole glory and what's more, in the duty bunk, which is close to the machines, where I can hear every irregular noise even in my sleep. But that's the point. I have to hear every disturbance and every command, because often it depends on seconds which the boat has to dive away.

"The L.I.[9] puts out a growl," says one of the watch officers when I'm asleep.

During the night, one of the two pistons of the cylinder cooling water pump on the starboard oil engine tore off and shattered. It then had to be exchanged for a spare

---

[9] L.I. means Leutnant-Ingenieur. Refers to Hans Fechter.

piston. This work is finished at five o'clock in the afternoon.

The weather has completely cleared up, it is calm and clear. The night is also relatively quiet. Only at four o'clock in the morning a dimmed vessel suddenly comes into view. We evade him on the surface. With the new day comes more sea, but without becoming uncomfortable. Towards evening it will be foggy. Since we are still too close to England's coast to be safe from surprises, the commander orders that we dive at eight o'clock. We stay down at night, only surfacing once at midnight to ventilate the boat well and get some fresh air back in because we are just beginning our journey and so need to economize on our oxygen and air purifier cartridges, not knowing if we're on our destination can receive new ones.

At four o'clock in the morning we surface. Both oil engines are turned on for driving and loading, because we didn't make much progress during the night and we want to reach our usual level as if possible. At the same time, the electricity consumed during the night is recharged into the battery.

After four hours the battery is full, so in order to use as little fuel as possible we continue our journey with the starboard oil engine at cruising speed.

The barograph is rising again; we are already looking forward to good travel weather. At midday, Black Rock is abeam — we've already covered a quarter of our entire way. Despite the relatively good weather, during the course of the afternoon one of the crew members, albeit a clerk recruit, who had been on land for years and had been taken along with his typewriter for only this voyage, became frantic with seasickness. Cold compresses, hot tea with and without rum, every remedy anyone knows of is tried on him and they succeed in calming him down so that he falls asleep.

The night passes without incident. The next morning the sea is calm and the weather is sunny.

At ten o'clock in the morning a sailing vessel comes into view.

We are now in such a favorable position we can stop him quietly without betraying our destination. We already worked here on the previous trip, and besides it is on our course, so we don't lose any time. We approach him and signal "send boat".

The order was followed on time, and shortly thereafter the sailing captain appeared on board with his ship's papers. It is the Russian schooner barque "Balzer" from Riga with a load of wood for England. Enemy ship with contraband goods.

The commander orders the captain to get into the boats with his crew. Having done this, we sail alongside the sailor and shoot a few holes in the waterline with our gun. We don't wait for it to sink, for we have some experience sinking ships and know that we have done a good job. At noon we see him sinking away on the horizon.[10]

At one o'clock another sailing vessel is reported, who can also be approached without losing time. Again we approach him and give our signal "Send boat". This time it's the Norwegian bark "Morna" from Stavanger, but she has wheat on board for Queenstown. Neutral ship with banned goods enters martial law. After the crew got into

[10] The 343 ton Balzer was attacked on August 10, 1915.

the boats, we also sink the "Morna" by artillery fire. She sank after an hour.[11]

At coffee time, a third sailor comes into view, and everything goes back to schedule: at four fifty minutes, the French three masted bark "Franxois" from Nantes sinks with wheat from Portland Oregon to Queenstown. We actually felt sorry for the crew on these ships, because they had all been at sea for fifty or seventy days, and now fate overtook them a day's journey from their destination.[12]

The next day we entered the Bay of Biscay, which is notorious for its bad weather. But we seem to have caught one of the three good days she's supposed to have in the year, as there's just a long flat Atlantic swell.

My engines also make me a happy day, because they run their daily workload perfectly. In these waters we can even close the dive tanks during the day. Diving will then take a little longer, but there are no surprises to worry about.

At night we try in vain to establish radio contact with U34. So we're slowly getting used to the idea that it turned around, as a result of the fault in one of its couplings, which it reported to us at Helgoland to repair it safely at home.

Cape Finisterre is abeam at ten seventeen in the evening.

The next morning the weather is so nice that we first open the front and later also the rear torpedo hatch, close all the

---

[11] The crew of the 1877 built, 1512 ton Morna was given ten minutes to abandon ship and reached the Irish coast the next morning with no casualties.

[12] The 21 man crew of the 1900 built, 2212ton three masted barque François were able to reach land the next day. The vessel had departed Portland, Oregon in March 1915 and endured rounding Cape Horn on this final voyage.

others and ventilate the boat well with the oil engines, which need a large amount of air to function.

In the course of the morning we are at Porto. In the absence of any other pastime, this important event must be celebrated with a glass of port. In general, either nothing is drunk on a U-boat or only very little. The first was the case for us. The result of this one glass was a leaden tiredness, which settled on the off-duty watch participants and soon united everyone in slumber. Well, we could afford it due to the nice, sunny and clear weather.

You're always tired on board, and if you have time, you sleep.

The sunny, warm weather and the clear water also tempted to swim. On the spur of the moment I got into my bathing suit and clung to the lowered F.T. mast on top of the dive tank. It was delicious to let the salty water run down your back. But I had to hold on tightly so as not to be swept overboard by the current. But how wonderfully fresh I felt after the bath.

In the afternoon Lisbon is abeam and thus covered half of our route. This night is also slept through; neither the enemy nor the evil enemy disturbed my rest.

The next day we had to be under water for an hour and a half, because we met a steamer which we could not pass without being seen. We dived since we didn't attach much importance to early reports of our presence reaching Gibraltar.

We continue a few hours as slowly as we can, for we intend to pass Gibraltar by night, and if we continue at this rate we shall arrive too early tomorrow.

Towards evening the diving tanks are opened again, because it is only a short distance to the strait leading into the Mediterranean Sea. But this night also passes quietly. A couple of times we have to turn away from steamers, once at five in the morning we got too close to one in the dark and dive so as not to be noticed, but only for an hour and a half. As soon as we're up there, I get the message: "Gyro compass failed." That was exactly what was missing when we wanted to get through the narrow area today. But the shock was greater than the damage itself: a man had accidentally hit the switch, knocking out the power to the system. After switching it on it worked perfectly again.

My engines are functioning splendidly; revolutions increase from hour to hour, as if they too knew what was important. I usually sit in the Central and look down the narrow aisle between the engines. A couple of times I go into the engine room myself to "see if things are all right; but basically all I have to do is look at the faces of my watch. I can read people's inner moods like in a mirror. They know today they have to show what they can do, and I can totally rely on them: they are more capable than the most capable.

You have to imagine what that means: waking six hours, sleeping six hours, waking six hours, sleeping six hours, and that for days and weeks. In these hours, however, there is no sign of fatigue: they feel the excitement of the situation, in which it is not so important to act as to be absolutely ready at every moment, and they wait for what is to come.

After a successful breakthrough, despite the slight fog, we can already see a good part of the Mediterranean, which now lies wide and clear in front of us.

The voyage through the straits is successful—and a strange coincidence means that today is the commander's birthday.[13]

So it's a double feast day.

At ten o'clock a small vessel, apparently a torpedo boat, comes into view, we dive.

After an hour we are back up and drive and load with both oil engines until about noon, then the battery is full and cruising is continued using the port oil engine in clear, sunny weather that comes up in the afternoon.

At ten o'clock in the evening another torpedo boat comes into view, in front of which we dive and stay down until midnight because we don't know whether it saw us. It's terrible not being allowed to attack anything, having to avoid everything in order to carry out our task of making a surprise appearance in the eastern Mediterranean.

The next day no ship bothered us. Nothing is in sight; we have time to get used to the southern climate. We've had a temperature of forty degrees Celsius in the engine rooms for about eight days, and it hasn't dropped below thirty degrees in the living quarters either.

And it's wet in the boat – unbelievable! Everything is moldy, books, clothes, linen, bread, boots—everything is covered with green mushrooms. It's particularly warm tonight. Thirty-five degrees Celsius is the minimum reached. Sleeping, be it during the day or at night, is out of the question at this temperature; you toss and turn restlessly from side to side. One imagines one is asleep and hears every sound, every sound, especially in my beloved watch bunk, which I always use to sleep on during surface

---

[13] August 16th is the date of the crossing/birthday.

cruising, because when we're underwater I have to be awake all the time.

Luckily the next day brought wind and with it some cooling. In the morning we got caught in a hurricane-like thunderstorm. Initially cloud banks rose, displaying all colors from the deepest black-blue to the palest green. They were closing in at great speed, and a wind broke so fierce and raging that the sea did not have time to follow it. The water was torn loose in shreds and lay flat over the sea like a veil. Add to that hail and rain — no one on watch could stay on the unprotected tower. You couldn't see five meters ahead if you could even open your eyes in the driving force of the rain. But as quickly as the weather came, it went just as quickly, and as I said, it had its good side, because it stayed a little cooler than usual until the evening.

The next few days passed quietly and peacefully.

During the day we sail with open hatches because it's already forty-five degrees Celsius on deck in the sun; the diving tanks are closed and every steamer is avoided. At night the hatches are closed and the diving tanks are opened. It was a quiet, peaceful voyage.

The only change was the annoyance of every steamer that had to be avoided; but what good was it, an order is an order!

Finally, at the penultimate position, the sedate cruising becomes too boring for us. There is still more than enough fuel, so both oil engines are "Full Speed Ahead" toward our destination to receive land stations. U35 drives as if it were competing with a torpedo boat - and at three o'clock in the afternoon we are standing in front of our destination and trying to maintain radio-telegraphic contact with the land stations. But heavy storm clouds hang in the mountains and no station responds. We then try to register with our signal lamp, but these signals don't reach their destination either. In addition, the weather is getting worse and worse.

What to do? Entering an unknown port that is closed during wartime without prior notice seems too risky. We decide to wait for the next day and then try our luck again. In order not to be disturbed by friend or foe at night, we dive at ten o'clock in the evening and stay under water until the next morning.

At eight in the morning we finally manage to establish the wireless connection. At one o'clock an allied power torpedo boat appears and guides us in. We are greeted with jubilation, and our achievement, which we ourselves regarded as a leisurely peace journey, is marveled at and admired.

But our astonishment is greatest when we are told that a torpedo boat has just set sail to pilot a second submarine. We ask, wait – and promptly an hour later, U34 moors next to us. And now the mystery of Gibraltar is explained. Since the German coast at ... we had moved ahead through our surface cruising and when near the English coast we had drawn the entire guard on us, so U34 could pass undisturbed and thus finally overtook us. In the Mediterranean we must have always been in the immediate vicinity, but out of sight. We had then overtaken our comrade through the fast driving the previous few days, a lead we had lost again by waiting outside the port. In any case, the boats that went out together arrived at their destination within an hour's difference.

Alarm Stations in U35

The Central.

For all diving maneuvers and other occasions when the boat has to be clear in order to disappear immediately into the depths, my place, the engineer's place, is in the control room. This room is immediately below the tower and forms the brain of the boat with all its facilities and the tower. You have to think of it like a big lager barrel with a flat bottom in it and three circular holes cut in the walls and overhead. The upper hole leads to the tower, the front one to the living quarters, or rather corridors, because they are as wide as the corridor between two rows of seats in an electric train car or a railway compartment; the rear hole leads to the engine rooms. The size of these holes is quite modest, a little over a three quarters of a meter in diameter, but we U-Boat men braggingly call something like this a door, and we get through it just as easily as a mere mortal would through the door to his own room.

The walls next to these holes are not smooth and shiny, on the contrary. Every spot is used, and there are a thousand nooks and crannies, like in a fully packed moving van. Two mighty columns rise in the middle from the ceiling to the floor, as if they wanted to carry the weight of the tower; but that is not the case at all, because they are just thin tubes in which contain periscopes when they are collected in, “retracted,” as the knowledgeable says. A lot of other things are contained by them: water levels, barometers, thermometers, humidity meters, speaking tubes, air testing devices, etc. What is most striking is the large mouthpiece, into which the commander's speaking tubes opens, through which he calls his orders down from the tower to the Central. It is kept wide, and during attacks etc. people like to listen to it in order to hear the commander's monologues and have an idea of an attack's prospects. When necessary, I even pushed in with my whole head so as not to miss a word or waste time asking questions, because sometimes seconds matter for us.

The equipment in the Central leaves a passageway on either side of the periscope shafts, but two men can only squeeze by each other through by keeping sideways. On one side there are hand wheels for depth control; in many cases they are already occupied in critical situations during surfaced maneuvers. So there is no more space on the side, because among these hand wheels also include much equipment, such as the depth pressure gauge, which shows the depth the boat is at, the water bubble, which indicate the boat's inclination, rudder position indicators for depth rudders and vertical rudders, screw shafts rotation indicators, coupling devices, etc. The space under the hand wheels is occupied by a large air pump, a gyrocompass system, fuse boxes, valves and pipes. Further along the front edge are large starter motors for the machinery with their associated electrical apparatus; There are also entire seacock and valve arrangements that are about as tall as a man - and a similar complex is opposite on the other side. The whole thing is, so to speak, the parade ground for my non-commissioned officers and men when diving. Commands are executed there by both, only with the difference that on a real parade ground a move that doesn't work can be done again, whereas here it has to work absolutely at the initial instance because it can't be repeated without consequences for the boat.

Pipes and valves abound again below these protuberances. There is also an entire compressed air center with a compressor, drive motor and all the accessories, and yet there is still enough space that you can comfortably stand there with three people. Strictly speaking, it's not really comfortable because two of them have to stand a little crooked, otherwise they'll hit the overhead, where the pipe string runs into the next pipe string. Wide ventilation pipes, medium-strong mouthpieces and air lines, thin manometer and lubrication tubes, everything is wild and confused and only clear and understandable to the initiated.

But the best thing about the place is the periscope that is terminates here. Through this you can observe the progress of battle when proceeding surfaced and communicate what you see to the watch personnel at the engines and with whom you communicate with via signs through the opening in the rear compartment, because the noise the surface engines make, my good oil engines, is too loud to penetrate with words, we have to shout everything into each other's ears. This periscope, the eye of the submarine, is one of the most wonderful facilities. In the depths underwater, even in the brightest sunshine and the clearest water, you could see no more than thirty to forty meters. The deeper you dive, the lesser the visibility. It is clear nothing can be done with these short distances, for the ships which one sees on the horizon when surfaced would long have disappeared if one tried to approach underwater in a straight line within thirty or forty meters visibility. And this distance is too short to be able to fire a torpedo at the target. Your own boat would also be destroyed. Therefore the eye of the submarine, even when it is at depth, must protrude out of the water into the air, as a kind of extension of the eye. And that's what the periscope is for.

It is a tube a few meters long that is sealed against the water on the outside and can be moved up and down vertically. At the top is a lens and prism system that keeps it fully waterproof, and a corresponding one at the bottom.

The horizontal light rays, which come, for example, from a ship that was seen on water surface horizon, are bent vertically downwards by the upper optical system and brought back into a horizontal direction at the bottom in a second lens arrangement. At the lower end of the periscope there is an eyepiece through which the viewer can observe the part of the horizon at which the upper lens system is aimed, with any target on it.

The field of vision is very limited in terms of width and height. If you want to search the entire horizon, you turn the periscope around its own axis with a few movements, the entire circle moves past the commander's eye little by little. It is a difficult piece of work with the periscope that requires a lot of practice to find every object, no matter how small, such as the mastheads of a distant ship, because its optics do not magnify, but rather reproduce the image in its natural size, at the same time you have to estimate distances using the instrument and therefore any magnification would lead to false conclusions. If the boat is at depth, the periscope only protrudes a short distance from the water. When traveling surfaced, you are on the tower several meters above the water's surface, so to speak on a high tower. If you dive, to continue with the example, you have climbed down from the observation tower, are standing on the ground, and the panoramic view has become considerably smaller.

The lower the periscope is retracted, the smaller the panoramic view it provides. If you search the horizon exclusively submerged, you may not find a ship or a sail in your periscope. When you then surface, you are often relatively close to a ship that you have not previously seen with your limited view. In addition, there is the turbulent sea as hindrance. Even if there is only little swell, wave after wave rolls over the upper eyepiece and obscures the spotted ship, if only for a moment, but still disturbingly. If the weather is smooth, the periscope sticks out of the water like a pole and can then easily be spotted by the attacked ship in good time for it to have time to flee. Therefore, when attacking, one may only show the periscope as little as possible and must accept all the disadvantages associated with it.

Incidentally, when you go underwater, looking into the periscope gives you a wonderful picture. Wrapped in the blue of the sky you can see the whole boat as if it were floating in the ether, and only the pearls of pleasure, seaweed, jellyfish or other sea creatures gliding past make you realize that it is moving forward. For hours one could stare out into the bright water, the fine color tones changing like a fairy tale with the depth of the boat and with the time of day.

Alarm Stations in U35

Alarm Stations in U35

Inside.

There is no more space at the rear end of Central than at the front. The stairs or rather the ladder leads up into the tower. Woe to him who is not used to climbing them, he inevitably bruises his kneecaps and shins and bruises his skull at the top. But everything is habit — and for us these stairs were downright comfortable. There is also the main bilge pump, which is only rarely used, because, thank God, the boats are usually so tight that you have to wait a long time before you reach for them. Otherwise, every corner is used. A large control panel with countless bare switches hangs above the pump, and fuses line up next to fuses like a pretty pattern.

This is my area, my attack and diving station. The engines are further aft. I can also observe them from where I am and above all, hear them. That's most important. The sense of hearing tunes into the noise so wonderfully that one knows every sound. Just as the musician hears every wrong note in his orchestra, so the engineer hears every irregularity in the way his engine is running, and he can often identify the cause of the disturbance from afar from the nature of the sound.

For this reason I had pitched my alarm bunk right in front of the round opening leading out of the control room; from there I hear everything, even if I didn't see it, and many a mistake I heard before the watch saw it, for their ears were dulled by being there for hours in the midst of the noise, they tended only to listen what eye focused on.

My alarm bunk was my own invention. Originally the space had been used differently for lockers, holders and the like -

and on the first trips I only camped there on unused hammocks. Only when the opportunity arose did I have my bunk built in, in which I could often be found during the day. There was a box that contained my most necessary books, tables, charts, etc., and there was usually some kind of book inside as well.

There I sat, read, wrote, and slept. If I leaned forward, I could see through to the crew berthing compartment. Here, too, there are always two berths one above the other, only here the head end is next to the head end, and berth follows berth. The non-commissioned officers live further ahead. The arrangement of their compartment is almost the same as of the crew; only there are curtains in front of each bunk, and a narrow table can be set up in the middle of the passageway. At mealtimes, the upper berths are folded down and together with the lower berths form a comfortable bench with a backrest. More space to fill out crew lockers.

Then follows the deck officers' compartment. The enlisted and non-commissioned officers have lockers and bunks made of tin; everything here is made of wood, cleanly polished - and with its shiny fittings the room makes a very homely impression. While everyone else has a metal washbasin of their own, here everyone has their own washbasin that fold into the wall, similar to those you sometimes find on trains.

Then comes the commander's quarters. The narrow passage in the middle also leads through it. His bunk is on one side of the aisle, around which are the lockers. But not all are for his use. The officers mess crockery is kept here, as well as the medicine cabinet and the official books and papers cabinet. On the other side of the corridor is the commander's desk and in front of it a small upholstered bench. But above all the compartments, just as in the

Central, the circular overhead arches, so that it is highest in the middle. You can almost always stand up straight here if you don't measure more than 1.85 meters.

All of these compartments are separated from each other by thin sheet metal walls. The usual circular hole in the middle leads to the bow compartment, which are also the officers' living quarters. At the rear end, the toilet is sealed off on one side by thin sheet metal walls. Opposite it are two Lilliputian clothes lockers, about the size of a grandfather clock, then there is another bunk on each side. The starboard berth was actually my domain, but I rarely slept in it at sea because it was too far away the center of activity. Only when there was nothing going on upstairs and everything was in sight did the crooked bones in the too short alarm bunk straighten out in sleep; otherwise the second watch officer used it.

When you hear the word bunk, don't imagine a bed with fine white covers and if necessary cover yourself with a blanket, but rather a plain horsehair mattress on which you lie down with your clothes on because on a submarine everyone sleeps in their clothes, washes and shaves every eight days. We officers had been afforded the luxury of having a bucket with water and a towel and soap for washing our hands, but the water was changed every three days at the earliest. Only the cook was ordered to have such a pail for washing his hands.

There's more space forward, the bunks are spaced farther apart, because there must be a clear area to load torpedoes; so there is a table for four people and we have eaten at it several times above and below water and spent many a pleasant hour after the day's work – thanks to the kind gifts Germany gave us. The commander of a ship usually takes his meals separately from the other officers. On a U-Boot, the spatial conditions alone mean this practice is difficult or impossible to carry out. Therefore, when duty allowed, we often sat in fours at our clean white tablecloth, the honorary title of tablecloth however only applied to the first three days, and then only when the weather was nice and the plates and bowls did not upset due to sea conditions. The splendour ended even sooner. Our meals for the first three days consisted of fresh food; later our cook knew how to prepare a tasty meal from canned vegetables and meat. We shared lunch with the whole crew. The cramped spaces do not allow the cook to work with many pots and pans. In addition, since only electric cooking is used, this would cost too much electricity and shorten the duration of the trip, since the electricity has to be restored by using up fuel supply. In addition to the cook, we had a very skilful steward who provided us with delicious meals in the mornings and evenings.

Only the bread had its downsides, after eight days it was mostly moldy through and through. Then we resorted to hard bread, the ship's biscuits. But in the long run it will cost you your teeth. So I decided and had an oven made from the electric ovens, and now flour was taken with me instead of bread; this was kept in soldered metal boxes, and freshly baked goods were baked every two to three days.

This was roughly the picture of the compartments in which I spent almost two continuous years of my life. And I must say that despite all the harsh demands of duty and service, this life has been good and harmonious. Not just among the officers, but also among the crew. I never heard of any quarrels or arguments. And there were no complaints about too much work either - even if the duty required eighteen hours of work per day; it was just done, taken for granted by everyone. This resulted in a spirit that we could not have wished for better or stronger.

In the Aegean Sea.

After an eight-day stay that was as enjoyable as it was busy, we are ready to sail again and leave port on a Tuesday at two in the afternoon to look for our field of activity in the Aegean Sea. Another torpedo boat takes us out, but by three o'clock we are alone and on our own two feet.

We start this journey with big raisins in the sack[14], because the Aegean is said to be teeming with warships and transports.

The very next morning we meet a steamer with two funnels. We dive, approach it, but the two smokestacks are deceptive, it's a small sea tug, not worth a torpedo shot - so keep going!

At eight o'clock we surface and continue our march until the afternoon. Another similar vessel comes into view. This time we go too far down to get a shot, dive and are back up after two hours. From now on we will not be disturbed either that day or the following night. The weather is calm and fine as usual in the Mediterranean at this time of year.

The next morning, the helmsman, who is currently on bridge watch, interrupts our breakfast with an air-raid alert. No one but him saw the airplane - but what did it matter, the crew was alarmed, and so we had to carry out the diving maneuver to its conclusion.

Misfortunes never come singly; as soon as we are under, the front hydroplane fails. With some trimming back and forth using water and crew, the boat can be kept level.

[14] German expression meaning to have high hopes.

We'll stay down and I investigate the cause of the disturbance. A cable connection has come loose and is only making intermittent contact. The connection is quickly restored; the hydroplane is turned on and tried, it works perfectly, and the boat obeys.

At night we continue our march and at one o'clock we pass Kitheria by bright moonlight. It was a beautiful night, the moon blazing bright, clearly illuminating mountains and islands. Thoughts and memories come up—from days when the world was peaceful. But at the same time the consciousness awakens and duty switches on: one has a submarine under one's feet and the responsibility of not allowing moonlight thoughts and feelings arise from the outset.

The coming day begins well; scarcely does morning dawn when an aircraft appears, but this time a real one, and forces us under water. Diving in front of aircraft seems to bring bad luck, because now the port electric motor shaft bearing begins to run hot and can no longer be kept in operation for extended periods. This means we only have one electric motor available for underwater travel and can no longer develop our full speed. After surfacing, damage is immediately determined and repairs begin. The bottom of the motor has decreased by 1.25 millimeters. The bearing has to be examined. After disassembly it is found the bearing metal has scored. There is no reserve bearing, so a lining must be placed under the lower shell to bring the old bearing higher again and it can continue to be used.

My people work splendidly with the heavy pieces in spite of the heat prevailing in the boat. I feel like I'm in a Vulcan forge with the muscular figures, clad only in trousers, dripping with sweat, tackle and master the heavy castings like children's toys in the swaying boat.

Shortly after lunch two masts, two funnels of a warship come into view.

We dive and move in to attack. It is an Italian "Lybia"-class small cruiser. Unfortunately we cannot develop our full speed with our damaged motor; nevertheless the commander manages to bring the boat within shooting range. Everything is made ready; everyone eagerly awaits the moment when the torpedo leaves the tube with its gentle shaking.

We in the Central hear the commander's orders he gives to the torpedo room:

“First tube. Ready..."

But the "Go!"[15] died on his lips, for the cruiser turned, veered away from us, and so escaped our torpedo at the last moment.

When we want to surface at six o'clock, the commander again discovers through the periscope a small cruiser moving towards us. It's getting dark, but maybe we'll still succeed.

We stalk. Time seems endless to us since we see and hear nothing. The torpedoes have long since been made ready, but there is still no indication from the tower of the command that we have managed to get to him.

Then, suddenly, after almost two hours, the commander's clear voice: "First tube ready!"

"It’s Ready."

"First Tube — Ready! - Go!"

---

[15] “Los!” in the original. The command to launch a torpedo.

The boat bucks back a little, and the torpedo rushes towards its victim, inaudible to the enemy. Everyone stands in the boat with a watch in their hand and counts the seconds until the detonation occurs. Fifty have already passed - not a sound to be heard - so - over! — It must have been pitch dark up there long ago; you can hardly see anything through the periscope. One tells oneself all the reasons why the torpedo missed its target; nevertheless, as always, the mood is a bit depressed after a missed shot.

We'll show up soon. Since we don't want to get into the tangle of islands with a damaged motor, we drive on at our slowest speed. Work on the bearing appears to be progressing rapidly. At midnight we have to dive again in front of a dimmed vessel whose character we cannot make out at night for the sky is cloudy. At dawn the starboard propeller is disengaged and the battery charged. At 7.30 the repairs to the port side bearing are finished, it is immediately tested and cruising commences. It's holding up well at first, but by midday I'm already certain continuous operation with it isn't feasible. So we disassemble it again and work on it. This ends at five o'clock in the afternoon.

The next day, a Sunday, sees us in the middle of the Cyclades island sea: with the rising sun we can see Naxos, Paros, Mykraio and Nicaria. After the cold night, the islands lie clear in front of us, gilded red by the sun's rays. Then a light haze rises from the sea and hides them from our sight again, only around noon do they become visible again. The sun beats down on us almost vertically, so that today it is cooler inside the boat than on deck; in the berthing compartments we had thirty to thirty-two degrees Celsius, at the engines forty to forty-five degrees Celsius. During the day there is no vessel in sight.

Our next destination is Mudros Bay on Lemnos. The night becomes thick and invisible. When at eight o'clock a dimmed steamer was suddenly noticed quite close to us, we preferred to wait underwater for moon rise. At midnight we emerge again and continue our journey on the surface. In the morning we have to dive in front of a steamer again to avoid being seen; because we want to have a go at the warships and not scare them away as they approach, since we are close to their assembly point.

Surfacing at five o'clock, Mount Athos and the islands of Lemnos and Strati are in sight. We set course for Lemnos.

A cloud of smoke is reported to the commander at noon. We dive and head for it with the utmost at full speed, but a French cruiser, "Jules Ferry" or "Victor Hugo", passes us only far, far from us. We can't get a shot.

We use the night for a cruise to the Gulf of Salonika, which proceeds calmly and without disturbance. In the morning at sunrise we have the fire of Kassandra to starboard, Mount Olympus, Ossa and Pelion to port amid short heavy seas. Old secondary memories are awakened here, where one rises and goes to sleep with classical antiquity; but troop landings don't seem to be going on yet, because we don't meet any vessels.

At eleven o'clock we sighted a submarine on the port side astern. Only the tower can be seen, which appears to have an aerial mast at the astern; some claim that it is a sailing vessel. A light, fine smoke rises behind the tower and now all doubts are resolved. It's a submarine. The silhouettes in Weyer[16] are eagerly browsed through to find one that has a thick air mast on the aft edge tower. We finally discovered

[16] Weyler's Warships. The printed annual worldwide warship database.

one that roughly fits. So it must be one of the two Greek submarines.

While searching, two steamers came into view ahead. One lies there stopped and the other drives up to him. We intend to investigate him. We dive and look at the two fellows underwater. The first is a small Greek coastal steamer; the commander cannot recognize the flag of the second. He thinks it is an enemy guard vessel, but it could also be neutral. In any case, in his opinion it's not worth a torpedo. So, in order not to violate neutrality, we drive past them underwater, emerge out of their sight and continue further into the bay as far as Kara Burun without finding any suitable object for our torpedoes.

At four o'clock in the afternoon we turn around and drive back half the way we came. After a good hour we meet the security vessel from the morning again and dive in front of it. When we surface, it is still close to us and Morses with light signals; which is not recognizable in the prevailing darkness. So we dive again and stay down for another hour. It is pitch black and not very visible. In order not to be surprised, we go underwater again until around midnight. Then the stars came out and after another unsuccessful day we marched back to Strati.

In the morning we are thirty nautical miles from Strati. We stay where we are and wait for things to come. But they don't come. Nothing happens. It's getting dull; this waiting for successes to come, and the mood is getting worse by the day. At two o'clock a small steamer comes into view, similar to the one we met yesterday in the Gulf of Salonika. The commander, however, held on tightly to his torpedoes. These objects are too worthless for such an expensive projectile, and we dive lest we chase away nobler game.

As it gets dark we see searchlights pointing in the direction of Mudros Bay. It is a cloudy and overcast sky. So we'll stay underwater until midnight. After surfacing, we drive closer to Lemnos and charge at the same time. Two hours later, a darkened vessel suddenly appears close to us and we go under. It seems as if traffic only occurs at night at this point. At four o'clock we come back up topside and spend a boring and wasted morning in the same place as yesterday. Shortly after the midday meal, a cloud of smoke comes into view. We dive and then, having lost sight of them again, we drive underwater close to and into Mudros Bay. Nothing can be seen in the outer bay, and only a little smoke rises in the inner bay. So nothing there either. We are close to despair. We've been underway for ten days now and haven't had any success yet.

We submerge at three o'clock and return. About a quarter of an hour had passed when a steamer with four funnels suddenly came out from behind the next cape and entered Mudros Bay at high speed, passing directly past the place where we were. We curse—but we can't help it: we can't get him because we can't go into the bay as we saw that there were nets in front of it.

So back again.

An hour later our misfortune is repeated: a four-funnel steamer, just as big as the first, maybe even the same, puts out of the bay without us being able to harm it in the slightest.

I relate all these trivial details to show the strain on nerves in a submarine when nothing is happening—or rather, when it isn't.

The constant driving without success, this searching without result is about the worst thing there is for the crew. Everyone feels a dull resentment and has no one and nothing to vent it on. So everyone is beginning to get nervous and upset. One walks restlessly and sleeplessly through the boat. Everyone, people and things, looks wrong to or at you. Sometimes one complains here, sometimes there, literally looking for an incident and thus spoils the mood for oneself and for others even more.

Of course, the lack of sleep contributes a lot to this. After all, people can exist without sleep, but then they need variety, events, and successes. On your feet day and night and nothing but mishaps without results, that makes the tiredness feel doubly difficult. When I was able to sleep for a few hours south of Strati, I felt like a new person, faced the new day with new hopes and was even able to improve the mood of my people a little.

Such days require a lot of self-control, especially for the commander. He knows full well how grateful the crew would be for even the smallest success, but he has to hold himself and his people back and keep his goal firmly in mind. Small successes can be had every day, but this automatically means the big ones are no longer possible. Small steamers can be encountered anywhere - but one will be sunk - and the large ones, which are still a possibility, are guaranteed to be scared away - the hunt is disrupted.

On such journeys, technical incidents are almost a relief. It's a change when the boat suddenly makes a "thump" — after a few seconds, again: thump — and you realize that it's touching ground. "Go shallower!" I call to depth control. Both hydroplanes are already up.

Now the commander is in the control room, he has been woken up a bit rudely.

"Grounding!" is reported to him.

"Go to ten meters!" he orders, climbs into the tower, takes out the periscope and looks around again to see if luck is finally going to be more favourable.

It's even more uncomfortable when the bearing which has already been repaired goes to hell again, because the armature of the electric motor has now even lowered by three millimeters and the shaft coupling connection is already rubbing in the coupling. There is nothing left but to place the lower shell of the bearing so high that the bearing is in the correct position again. However, the port side is out of operation again for the duration of the repair, and all the restrictions mentioned are the result. Of course, just then a good-sized steamer comes into view, which we cannot reach now with our reduced speed. You get a sense of gallows humor — and ask yourself, half amused: what comes next?

When we show up after two hours, we know: the rear auxiliary bilge pump stops when loading. The guide control blocks have been softened by the oily bilge water and the valves are stuck. A small thing, certainly: but given the sensible complexity of a submarine, every single piece of machinery on board is of vital importance. If one fails somewhere, a hundred new difficulties immediately arise: one always follows from the other. Now the rear bilges cannot be emptied. The water collects in them, boat trim becomes messy, and depth control becomes uncomfortable due to the back and forth free flowing water. All the more uncomfortable when the examination turns out that the sliders are broken. Reserve slides have been used up before, and we can't get new ones in the port. Now we have to make them ourselves using our primitive means, hammer, chisel, file and saw.

I'm in the middle of full activity at the back with guidance and instructions when the call: "Watch to diving station!"

echoes through the boat again. I jump to the Central and flood - but the boat doesn't want to submerge. There's quite a swell up there, and it sticks, as they say. Both hydroplanes are down, one of the electric motors is already running at high speed, but the boat doesn't want to go down. Finally, it slowly tilts forward and begins to dive, but now the accumulated bilge water also shoots forward, and the forward momentum rapidly increases. All hands aft. Hydroplanes are reversed but the boat cannot be stopped. Emergency speed ahead – it continues diving.

The boat is still diving, finally it holds.

"To diving stations!" The men come back from aft, staggering and sliding, whereas previously they had to laboriously climb uphill.

The boat is now rising again just as quickly as it dived before.

The boat is leveled and becomes aft-heavy. Gurgling, the bilge water shoots astern again.

“All hands forward!” Only the most important remain at their stations and now the stern settles again. Once again, all measures are of no use, and compressed air is brought in as a last resort.

We are down aft again when the boat finally stops and levels. The attempt to vent the rear tanks in time also fails, and now we are up on the service.

The steamer we dived from passes five hundred yards behind our stern. It's a hospital ship and didn't see our small boat that just peeks out with its tower.

Such a fore or aft down angle at a certain inclination can be disastrous for the boat. Because if the battery acid overflows it can cause a short circuit and fire which destroys the battery, and without battery power you can't drive underwater, so you're hopeless at the mercy of a superior opponent without the possibility of diving.

With cunning and trickery I manage to get the water out of the bilges. It takes longer than with the pump we are currently working on, but it works, and when three hours later the call "Watch to diving stations!" sounds again, my boat dives calmly and safely like a duck.

We were fortunate, for shortly after midnight we suddenly had five destroyers ahead and had to dive in front them as quickly as possible in order to remain underwater until dawn.

The excellent work of our men contributed a lot to the quick repair of the damage. You have to imagine what it means to produce precision pieces from simple flat iron in a vise, without a workbench and a milling machine, in such a short time as was necessary in this case. Admittedly, it must be said that our nightly ascent and descent also spurred them to hurry; they too, from what I saw and heard, seemed to attach little importance to repeating this maneuver.

Around midday it seems that we are about to achieve something: the masts of a vessel appear on the horizon. We go underwater and approach it. But hope remains, because when we reach it, it turns out that they are a few ancient ships of the line, perhaps from Nelson's time, one of which is towing the other as if in a funeral procession. Although the view is once again poor, we still have to laugh at the sight of these old men of war, who, together with the tugboat accompanying them, were not worth a torpedo. As a replacement, in the evening all the electric cooking pots suddenly fell apart and our supply of alcohol was also exhausted, so we could no longer solder them on and repair the resistors inside. Gloomy food prospects add to everything else before us; but here, too, a solution is reached: with some devices that are actually intended for completely different purposes, it is possible to penetrate into the interior of the important pots and get them running again. Without that we wouldn't have had a hot meal on this trip, and our cook was already shivering at the thought of what he would have heard if he had had to put us on cold food.

The next day around noon we have Palmos abeam with the monastery of St. John shining brightly in the sun. It's a gorgeous sight. The monastery lies peacefully at the top of the island's highest point: its buildings greet us white in the sun. With the glass you can even see all the details clearly. To compensate for the joy we feel in this piece of beauty in the world, a few hours later it turns out that there must be a leak in an oil cooler on the port side, because our oil supplies were dwindling rapidly when this side was in operation. Our wake also shows a tell-tale oil trail. So take it out! We immediately start replacing the piping system, finish this work after twelve hours and now openly ask each other the question: What will the next breakdown bring?

But when our supplies were already running low and we were preparing to return home, something else was to happen: Heaven granted us at least a little compensation for our bad luck.

After a quiet night, used for a good night's sleep, a steamer came in sight at nine o'clock in the morning; we run after him and an hour later fire a halting shot with our gun, at the same time the signal is made: "Send boat!" The steamer obediently stops and sets a boat ashore. It is a Frenchman: the captain comes to us with his ship's papers. They show the steamer belonged to Austria before the war's commencement and now she carries coal for the English Admiralty coming from Famagusta to Bizerte. We'll keep the captain on board and send the chief watch officer across with two men, to examine the ship more closely. He sent the crew into the boats and brought us important news about the steamer routes.

At one o'clock we approached the steamer, which had been abandoned by the crew, and fired six shots into the waterline, whereupon it sank half an hour later. We tow the boats out of gratitude for our first success, which we achieved after two and a half weeks of unsuccessful driving back and forth, and towards evening we sailed them within ten nautical miles of Crete.[17]

Here we throw them off and continue our journey using our newfound experiences.

Bad weather sets in during the night; our course takes us across the sea, so that we are only wobbling insanely. Even during the day the weather held up—we couldn't get a shot again.

The next morning dawns grey. The watch is constantly trying to penetrate the lighter and brighter shadows of the night with their glasses.

"Lieutenant, there, four points to port, there's something!"

My glasses are aimed there.

"Report to the commander: Four points to port - a steamer," calls the watch officer in the tower.

Less than three minutes pass before the commander's head appears in the tower hatch.

---

[17] The date is September 17, 1915 and the vessel is the Ravitailler built in 1901 as the Austro-Hungarian flagged Gradac. 2844tons.

Now it's clearly visible: it's a steamer that's approaching us at an angle. The watch is already climbing down from the tower to the Central. We dive. We'll check him out underwater to see if he's armed. But there is no gun to be seen, so the commander decides to attack him on the surface.

"Surface, gun crew on deck!"

The first warning shot rumbles through the early morning. The steamer turns and tries to escape, but we are too close for it to be able to. The second shot, a hit in the deck structures, teaches him about it.

But he still hopes to escape his fate by fleeing. Shot after shot is sent after him until he realizes escape is impossible. He stops, hoists the English flag and starts launching his boats. Only now do we notice some people are already lying in the water and are trying to get into the abandoned boats or are climbing back on board on the boat falls. Three smaller boats reach the water very quickly and hastily move away. Gradually, more and more people appear on deck; the number of boats launched is far too small to accommodate them all. All the people are wearing yellow khakis, all are uniformly dressed—they are Indian troops for the western front.

Now all the boats are launched, all overcrowded, they are working hard in the rough sea. There is only one white man in each boat, and the Indians apparently cannot operate an oar. Here and there a boat capsizes; it is a sad sight - one would like to save and must not. Because every one of them who comes ashore has to fight Germans afterwards.

Now the deck is almost empty and there are no more boats on board. Only a few Indians are still huddled together on the stern and waving white handkerchiefs. They last us, but what's the use? Quickly fire a few shots into the forecastle, because the sooner this torment is ended, the better it is. The forecastle lists lower and deeper, and the bow enters the water.

But what is that: Instead of the individual people who are now standing on the stern, hatches and doors are suddenly spewing out people after people.

More and more masses swell up. The English must have kept them locked in the lower decks, and only now that water is flooding in have they broke out of their prison. Some jump overboard with beams; most climb down the hanging man-ropes and boat-falls and hang there, one on top of the other, like gigantic fantastic bunches of grapes. Horrified, we turn away to escape this picture of horror. One feels like a judge who, with a cold heart, has to execute a sentence here on the open sea, where one is meant to help one another.

As we drive away we only see the three small boats sailing away with the English in them who led those poor, unsuspecting Indians to their deaths. We leave them unmolested—out of a feeling of disgust at men who, in the most seaworthy boats, first abandoned their lost ship and the troops entrusted to them. And in spite of the success, no happy mood arises - the pictures remain until new work brings new ones and the hard commandment of war again suppresses feelings.[18]

---

[18] The 1905 built British transport ship Ramazan, 3,477 grt, was sunk on September 19 . The ship was on a voyage from Alexandria to Mudros carrying military supplies and some 380 Indian troops, (75 of whom survived the sinking, together with 28 of the crew).

U35's War Diary relates:
19.9.15, 4h30-5h20 in the morning, West of Crete, wind WNW 2, clear, visible.

Around noon a large steamer with two funnels comes into view. We try to attack him underwater, but we can't get to him, and a similar occurrence in the evening hours when another steamer comes into view. After diving, it is almost impossible to find him through the periscope in the twilight, let alone a chance of hitting him safely with a torpedo. In addition, it was no longer possible to tell whether he was neutral or hostile.

The commandant is happy to have finally found a place worth staying at and wants to make himself at home here. But now I have to put a spanner in the works for him. There's still enough fuel for three days, and we're still at least two days' march from the port. As much as it pains us, we have to let go of the prospect of success that we have finally achieved.

So we finally start our journey home and have already covered a good bit of the journey by the next morning, when another steamer comes into sight. We dive and make

---

Dived at dawn in front of dimmed vessel and drove to attack. Steamer lies so high out of the water that torpedo may pass under him. Therefore surfaced and fired a shot across the bow. Austrian flag shown. Steamer does not show flag, turns away and tries to escape; therefore opened gunfire on her. The steamer stops after about 6 to 8 observed hits and launches boats. By 7 a.m. 7 boats are launched fully manned; it is determined that it is an Indian troop transport. Men wear brown and yellow turbans, some tropical hats. When the last boat had set off at 7 a.m., the steamer was hit a few times at the forward hold waterline. A large number of men suddenly came running up on deck from the forecastle as the ship began to sink, but they did nothing to save themselves. The actual crew of the steamer had already disembarked and it seemed as if the troops which just appeared had been locked below deck. In total, the transport must have been about 500 men strong. The number of lifeboats on the ship was far from sufficient. Unfortunately, it was not possible to think of rescuing by submarine because of the excessive number of men. The ship was the English steamer "Ramazan" from Liverpool (2,200 GRT). It had black funnel with white ring and white and red flag underneath. On the forward decks above the forward holds, as well as on the middle deck, wooden sheds - apparently living quarters - had been built.

an attack, since the steamer has the insignia of a transporter. We make our approach at seven nineteen; I'm standing at the hydroplane station myself. The torpedo leaves the tube with the familiar, barely noticeable shock. He's barely gone for a few seconds when the commander up in the tower begins to fuss because he sees the torpedo take a turn in its path and miss its target. Shortly afterwards comes the command:

“Surface! Gun crew on deck!"

After the warning shot and a little more persuasion, the steamer agreed to stop and launch boats onto the water. It was an English transport, displacing 2,700 tons. [19]

When the crew is on board, the watch officer and two men are sent over with our dinghy to inspect the ship. The steamer is in the service of the English Admiralty and carries general cargo, mainly machine parts, from Mudros to Malta. Our people are still investigating when a second steamer comes into view. Quickly everyone back on board, use the dinghy, shoot a few holes in the first ship with the cannon and then dive. The second steamer, although it must have noticed what we were doing, comes up unconcerned; it is a hospital ship, picks up the boats from the sunken steamer and then quietly sails on.

We reappear and hit the Englishman again with the gun, because the first holes he received sucked in too little water, so he was sinking too slowly for us. No sooner have we finished this second, improved edition than another

---

[19] The 1906 built 4306 ton Linkmoor was carrying coal from Lemnos to Malta was sunk on Sept 20, 1915. The Germans found confidential instructions date June 1915 from Malta's Admiral Superintendent detailing actions to be taken when confronted by a submarine. These and other captured documents provided Germany with reasons to treat all defensively armed merchant vessels as warships leading to unrestricted submarine warfare.

steamer comes into view, and again we expect it underwater. Its front mast carries a yard and it has two smokestacks, so a typical Messageries Maritimes steamer. But he must have smelled a rat when he saw the stopped comrade lying there, so he gave this spot a wide berth such that we couldn't get within shooting range.

We continue our journey home — and now it's as if all the bad luck we've already experienced on this trip is about to come over me in a concentrated form. A radiator is leaking on starboard, on port side too, the cooling water pump is out of order and is no longer pumping, the rear auxiliary bilge pump broke for the second time, the periscope lift only works intermittently.

A rivet on the hull is leaking and seawater is dripping onto the battery when underwater. You don't know where to start or where to end. It's been going on for so long and it will probably continue like this for another twenty-four hours. Anything that can be done will be properly repaired right away; where parts are lacking, a makeshift solution is created and it works.

The next morning at ten o'clock land is in sight, and at four o'clock we are safely back in port. The fuel supply we bring with us is minimal. We could have driven another twelve hours — so we didn't lose any time on this trip. And what else we brought with us was also valuable enough; because another boat was already waiting in the harbor ready to leave for our return, to take our experiences with them on the way. On the same day she put to sea and returned twelve days later with a booty of twelve steamers and one sailing ship with a total tonnage of fifty thousand gross registered tons. That was partly to our credit.[20]

---

[20] U35 returned to Cattaro on September 22. U33 departed afterwards and sank 12 vessels between Sept. 30 – Oct. 7, 1915.

The submarine that we met in the Gulf of Salonika was U21, which was working in the same area with the same success as us, and had seen some of the same steamers as us off Mudros Bay.

The air mast, however, with the thick head on top, which we saw on the U-boat in the Gulf of Salonika and which made us mistake it for a Greek one, the poorly mannered claim was a watch officer with an almost Bismarckian sized skull.

Alarm Stations in U35

The submarine trap "Melanie"

We spent some time out in the open water working again. It was toward morning; about six o'clock it was getting dark; at six thirty the first steamer is in sight. Both oil engines give their utmost, and soon we are in front of him, diving to have a good look at him underwater and, if he is armed, to sink him by torpedo shoot; for armed ships are considered ships of war and need not be warned prior to being attacked.

There is no sign of a gun. We let him gain a little head start anyway, then surface and put in a few shots across his bow. He stops and his crew goes into the boats. It is the English ore steamer "Southerland" from Newcastle, with 3,600 gross reg. tons, which has coal on board. We approach her and sink her with gunfire. She sinks at nine o'clock in the morning.[21]

Half an hour later, a second steamer comes into view. Again the oil-engines have be run at their utmost, but there is no prospect of putting us in front of him and attacking him underwater; therefore attempts are made to force him to stop with gunfire. We open fire, he turns away, and a short time later there's a flash over there too. He fires a stern gun.

The impact of his shells is still a good eight hundred to a thousand meters in front of our bow, so we hold on and let it come down to who can shoot better, he or us.

---

[21] The Southerland(correct spelling) was sunk on January 17, 1916.

Shot after shot rang out across the sea – he answered at the same pace.

During this pursuit another steamer comes into view, which, though no doubt watching the action, calmly maintains its course, which intersects ours. She is soon within firing range, and as we have seen the first steamer is a tough opponent, we suddenly turn to the second and take a few shots across her bow. He turns and runs away. We keep firing – and after a few shots he stops and hoists a relatively large and new Dutch flag.

We are now about four thousand meters from him and give our well-known signal: "Send boat!"

In a short time, a book is launched over there and is coming towards us.

We have time to contemplate his behavior until the boat arrives. Either he has a very good conscience and therefore boldly continued on his course, or he is not as harmless as he wants to appear. So: better safe than sorry. We'll first look at it underwater. We dive; the few thousand meters that separate us from it are quickly covered and we circle around it at a distance of a hundred meters. The periscope is completely extended and protrudes at least two meters above the surface. The commander[22] looks his eyes out, but can't see any guns. The deck is completely uncluttered. Nor are there any flaps visible on the bulwarks behind which guns could be hidden. A few men stand harmlessly on deck and watch the spectacle. The new large Dutch flag flies at the stern and the name “Melanie” is emblazoned beneath it.

[22] Kptlt. Lothar von Arnauld de la Perière assumed command on Nov.13, 1915.

We keep thinking. If it were a submarine trap in disguise, he would certainly shoot at the periscope now, because even an Englishman doesn't have that much nerve to see a periscope sticking out of the water next to him, at any moment to get a torpedo between his frames, and having the opportunity to fight back and yet not doing it. We come to the conclusion that he is probably really neutral after all, proceed underwater towards him and emerge a little to the side and aft of it.

"Gun crew on deck!"

The crew hurriedly climbed up, I stayed in the control room and, at the commander's command, brought the boat all the way to the surface by blowing out the tanks.

Suddenly, in the midst of the noise made by the fan, the shouting of orders, the rumble of the rushing gun crew, suddenly another, unknown, strange noise!

Whoosh! – Whoosh! It howls over the tower, and a hissing blast of air penetrates as far as the control room.

Immediately behind them follow the last of the gun crews, who could only have just arrived in the tower, no longer climbing and slowly but almost falling—thump—thump—thump, one after the other, all of a sudden they're down again. I yank the switch, interrupt everything, and without waiting for the commander's command, I yell, "Watch diving stations, quick flood!" Simultaneously, the engine telegraphs rattle, "Emergency Speed ahead!"

Amid the commotion the crew who were on deck are still falling, one after the other. It's crowded like at a fair.

"The dog is shooting," someone hisses through his teeth as he pushes by.

"All vents are open except for six and seven."

"Flood Six and Seven!" echoes the commander's voice from the tower.

"Flooding."

To do this, I tear open the bottom valve and flood whatever wants to go into the ballast tanks; just get underwater before before we receive a hit.

Now the last watch officer climbs down from the tower into the control room and takes over the depth control, and slowly, far too slowly, the boat moves down. We don't know if he's still shooting or if we just can't hear it because the turret hatch is closed and there's so much noise - there it sounds again: crack, crack, crack. The shells burst in the water next to the boat.

Anxious seconds pass, which seem endless - then a look at the depth gauge: it shows fifteen meters - we are safe.

"Go to twenty yards — steer the boat down!" The commander orders down from the tower. This is necessary because I made the boat much too heavy by flooding it; after about five minutes, however, it is controlled and completely in our hands. Now the questions start. What was? What is?

It went like this: as soon as we surfaced and waved to the steamer's boat, which was about six hundred yards away from him, coming towards us, he suddenly opened fire on us with one of several medium-caliber guns and machine guns, salvos, which luckily went over the tower. Everyone on deck and on the conning tower jumped back inside, only the officer on watch staying under cover of fire in the lee on deck. At any moment the second salvo would be fired, which would surely strike us at such a distance. From the deck he climbs back up to the tower and wants to get inside.[23]

But the tower hatch is already sealed, at the same time he sees our boat diving away. Without further ado, he tears open the tower hatch again, jumps into the tower and then slams the hatch behind him. So he got into the boat at the last moment and was the last to reach the Central. With the engines going at full power, our boat had shot ahead, and during the short time we were still visible was behind the steamer's dinghy. So he couldn't fire his second salvo, which would surely have been our undoing, without endangering his people in the boat. That had been our salvation.

"Go to ten meters!" suddenly comes an order from the commander from the tower. "Ready Tube Two!"

Great! We're going to attack! As soon as the steamer noticed our periscope, the shells crackled around us again. But it doesn't matter as long as it doesn't get struck. We finally snaked our way up to him.

"Tube Two Go!"

After a few seconds the detonation must follow. We wait and wait—but nothing can be heard except the sound of shells bursting.

"He outmaneuvered it!" calls the commander down from the tower.

Regrets on all sides, because we were all looking forward to a just punishment. Shoot another torpedo? The other now knows that we want to finish him off, he's made clever maneuvers so far - so it's very questionable whether the second hits better. And further: It is still early in the day. If

---

[23] Leutnant Otto Launburg

he gets more guards by radio, they can make life miserable enough for us and the battery is half empty. So it's better we give up and drive away with heavy hearts, without taking revenge.[24]

At three o'clock in the afternoon we are out of sight, surface, charge the battery and continue our journey.

But we are still sorry the Englishman outmaneuvered the torpedo.

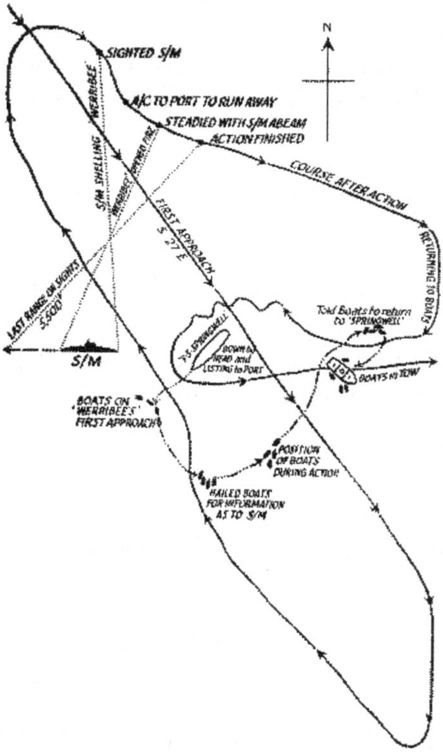

Approximate Movements of 'Werribee' in Action with Submarine on February 9, 1916.

---

[24] The vessel chased is the British steamer Springwell. The Q.ship engagement was with HMS Weribee on February 9, 1916. E. Keble Chatterton's "Q.Ships and Their Story", Sidgwick and London Ltd. London,1929.

The Sinking of the "Gallia"[25]

We are sitting warm in the bow compartment and are just about to make ourselves comfortable with a cup of coffee when a sailor sticks his head through the bulkhead door and calls out: "Kapitänleutnant! Report from the officer on watch: Clouds of smoke three points on the port side ahead."

"Hold on to it. I'll be right there."

As soon as he's gone, a second head shoves through the bulkhead.

"Kapitänleutnant! It's a big one, it has three funnels!"

Jumping up, we rush into the tower or to the Central in a single bunch.

As soon as I'm there, the order comes from above: "Watch to diving stations!" and our boat disappears from the scene without a trace.

"Level the boat. Both motors all ahead. Battery level?"

"Battery is full," I report back.

If only we could get a shot! We still have one torpedo, our last, and it would be a good catch at the end of the voyage.

---

[25] Events related in this chapter deal with October 4, 1916. Several cruises since the previous chapter are omitted. Of these many include Italian vessel sinkings. Germany was not at war with Italy until August 28, 1916 and would use the Austro-Hungarian flag when making surface attacks. (Note footnoted War Diary entry in Ramazan sinking indicating hoisting AH flag.)

Everyone is quiet and waiting. Everyone is also curious if we will succeed.

After a while the commander says down into the control room, "It doesn't come towards us and we can't get him."

A universal "Pity" is the response.

"What does he look like?" we ask back.

"Painted grey!"

How rotten—that means a transport not the usual hospital ship. A disgruntled silence falls on everyone. We're going at full speed: the battery is giving it its all - but it still doesn't help. Inside you're arguing with the good old boat that has brought us a lot of success and has never let us down when it mattered most. But what good does it do? You have to accept it.

"How fast is he going?" we ask again.

"I guess Eighteen to twenty knots!"

In this case there really isn't anything that can be done.

"It's already over!"...

Well, that's the final decision.

We're already thinking about the appearance again - suddenly a completely different note emanates from the tower:

"He turns, he turns!"

Everyone has new hope.

"He's zigzagging and turning toward us!" the commander's voice resounds happily. Every word is passed quietly from man to man around the boat, and anticipation shines in everyone's eyes again. And now the waiting begins again. The Commander peeks repeatedly—we expect updated bad news from the tower at any moment.

The last torpedo is already ready; it just awaits the signal to begin its deadly run.

Suddenly it sounds: "Ready Tube!"

"Ready! - Go!"

Now he's outside. The seconds are counted with bated breath.

"Twenty – thirty -- forty"

Boom -- Boom! It roars in the boat.

A loud hurrah! — It's a hit.

"I think I just blew off his screws, he just turned away again!" says the commander. "The water column was behind the stern."

"May I have a look?" — "Yes."

I jump up and look through the periscope.

The steamer is already sinking aft and has a slight list.

"Kapitänleutnant, he's going down!"

One by one the crew is sent into the tower. Stopped, the giant lies there with an increasing list to starboard and its aft-list increasing. Five minutes later I'm back upstairs and notice that he is much more down by the stern.

"He's going down for certain, Kapitänleutnant."

One after the other of our crew crowd towards the periscope, returning to their duty station with a beaming face, and there quietly report to their comrades what they had seen. It doesn't have to stay in the Central all the time either; once again I must make sure that it is sinking. I have waited for an opportune moment. The stern has submerged up to the bridge deck; the forecastle sticks out of the water and begins to sink slowly at first, then faster and faster".

Eighteen minutes have elapsed after the detonation when the seas close over the "Gallia".

Outside the sea is as smooth as glass — we count ten boats in the twilight. We do not believe that such a large ship sails entirely without escort, and so we continue underwater until it is quite dark. But no one bothers us— and as we found out later, the Gallia was actually making nineteen knots and unescorted.

Our hit had destroyed her radio station, so she couldn't call for help.

Three days later, the boats with the surviving remnant of the troops destined for the Salonika army landed in Sardinia.[26]

[26] The 1913 built Gallia, 14966 tons was armed as an auxiliary cruiser and used as a troop transport. The Gallia left Toulon on Oct. 3, 1916 unescorted for Thessaloniki carrying 1,650 French soldiers, 350 Serbian soldiers, 350 crew and artillery and munitions. Due to poor recordkeeping there were approximately 1200-1300 survivors and an estimated at 800-1400 dead. The captain went down with the ship.

How "La Provence" was sunk.[27]

We had been out on the transport route for a few days and had tried to catch this or that steamer. But they had all passed far from us.

It is past Noon. The sun shines down through a fine haze, but it is already warming despite the early season. The commander, the watch, everyone who wants to get some fresh air stands on the tower and smokes a pipe, cigar or cigarette. The moment must be taken advantage of, because in the previous days there was little time for anyone to come on deck and treat themselves to the pleasures of tobacco. A pleasant, tired midday mood hangs over everyone; the sun warms so beautifully. The splashes of the sea do not reach the tower, and the haze is annoying, because there are certainly a lot of steamers passing quite close by that we can't see; but we can't change that, and so everyone is happy to let the sun shine on their backs for a short time.

Suddenly the voice of the officer on watch rings out: "Get Below! Two points off starboard a steamer with two funnels. He's already gone," he then turns to the commander, who puts the glass to his eye.

Men hurriedly climb into the tower, many a recently lit cigar or cigarette flies outboard, followed by a curse.

I quickly peek again for the direction the steamer is supposed to be in, but can't find it and then disappear into the tower as well.

[27] The event in this chapter correspond to February 16, 1916.

As soon as I'm in the control room, the order comes: "Watch to diving stations!" In a moment it is done: the noise of the oil engines stops and electric motors start humming their song. The depth manometer slowly increases, and the wheels of the hydroplanes are tossed back and forth in a whirling motion.

"Ten meters," reports the head of depth control.

"Go to fifteen meters, trim the boat!" orders the commander/ after he has once again made sure through the periscope that the boat is heading towards the steamer on the correct course.

"Boat is trimmed!"

We approach stalking at high speed!

"First and second tube ready!"

That's quick; we must be close to shooting range.

It only lasts a few more minutes—then the commander's voice sounds again:

"First tube — ready! Go!"

The torpedo leaves the tube with a weak jolt, only gently shaking the boat.

Eyes cling tensely to their watch's seconds hand, everyone counts and waits.

"Thirty -- seconds -- forty" -- Nothing can be heard; everyone is stony still waiting to experience the detonation.

The seconds slowly crawl around in circles.

"Fifty -- sixty -- still nothing."

Did it miss? A curse slowly forms on lips.

Damn it's over! A big roast escaped us again. Disappointment is evident on everyone's faces, and not just quiet, but also loud criticism.

"Sir, did you see the wake?

"No, too much sea!"

In short, it definitely missed. What could it have been? Did the torpedo not run straight, did it jump out, or did it dive in front of the target? You consider all the possibilities — and yet you keep thinking: over — over.

Suddenly an unmistakable sound: Ping! Fum!

We look at each other as if we don't believe ourselves; but it is like this: it is a torpedo explosion and that means a hit.

"Congratulations on the master shot!"

"He must have been incredibly far away."

"What's he doing?" I ask upstairs.

"List to starboard and blowing off!"

"May the crew come into the tower?"

"Yes, but send the men up one by one!"

I'm the first up.

There lies the steamer and listing moreso. Clouds of white steam rise from the rear funnel, that's where the hit must be.

“Congratulations again, Kapitänleutnant; To hit him was an achievement."

One by one the men are sent to the periscope and, returning to their post, report what they have seen. But the joy does not last long — twenty minutes after the shot the steamer disappears, never to be seen again. We run off underwater and emerge an hour later. Nobody saw us. As a result, wild rumors ran in the French press: the Germans now had submarines without periscopes.

At first we didn't find out what the name of the steamer was—for us it was just a gray-painted transporter with two funnels and two masts.

It was not until we were in port that we learned that it was the French auxiliary cruiser "La Provence" with troops for Salonika, of which around 3,400 men drowned and only around 900 were saved.[28]

Six weeks later we were back on the same spot. The lookout spotted a floating life jacket: it must have been a submarine at work. On approaching a name was noticed on it, and as we were curious as to the ship's name which someone else sunk, and as there was nothing in sight, so we wasted no time, we set about fishing it. We pulled up and lifted it out of the water.

What was the name of the ship on it?

"La Provence".

A peculiar feeling crept over us.

---

[28] The 1905 built La Provence, 13,753 tons was armed as an auxiliary cruiser and used as a troop transport. The ship departed Toulon on February 23, 1916 with 2000 French soldiers, 400 crew and 200 horses and mules. About 870 men survived the sinking. Initial French reports indicated 3130 fatalities. The captain went down with the ship.

The Trip to Cartagena.[29]

We were ready to leave. Then it was announced: "Wait for further orders!" General astonishment; no one suspects the reason for the delay - and the grizzled combat veteran game of guessing and fabricating is played with zeal. A shipment arrives. Once on board they all have medicine in them, new assumptions, new declarations. But when saluting ammunition was bought, there was no end to the wondering, and the wildest rumors found apostles and adherents. Someone very clever wanted to be heard that due to new orders ships were only be stopped with a blank gun salute and forced to stop - this nonsense too was eagerly discussed and talked about. Clothing inspection is called for; articles placed in order and, despite previous habit, the dress outfit was also brought on board. We officers soon knew what was going on, but it remained a secret to the rest of the crew - and so the guessing went on happily and cheerfully.

Then came the day of departure. Everyone was anxious to learn something and hoping for it when the boat proceeded to the open sea. - Even then the commander still said nothing, and only when we were certain our voyage was to succeed did he have the crew line up and tell them the trip's purpose.

In a short speech he explained the Emperor himself had sent us to Cartagena with a letter in his own hand to the King of Spain, in which he expressed his thanks for the good treatment of the Cameroon soldiers who were interned in Spain, and for whom we had a shipment of

[29] Events in this chapter correspond to June 1916.

medicines bound for a Spanish port. He hoped we would succeed in the enterprise, etc., etc.

The voyage was straightforward and peaceful, and one fine morning at five o'clock Central European Time we stood in our boat at the entrance to Cartagena. Tired and sleepy – its three o'clock in the morning according to the local time – a rowing boat comes alongside: the pilot. He is amazed when he sees our vessel and tries to make himself understood, but none of us understand Spanish.

Finally he stands on deck and tries to make it clear to us by signs which place we should go to. But we find our way around and moor without further ado next to the German steamer "Roma", which is in the harbour.

When we have moored, its crew, which consists only of ship's officers, is already on their feet. First the work begins, medicines are transferred on board and inquiries are made if there are any enemy ships in the harbour; because we still find it unclear how long we can stay there. We knew from the start we would not exceed the prescribed twenty-four hours; we weren't in distress, didn't need any equipment and just wanted to do our job.[30]

Enemy warships were not in port. However the day before, in the afternoon, the French auxiliary cruiser "Colibri" left Cartagena. It had been lying here for days and apparently expected our arrival - but gave up this objective just in time as it was unprofitable. So we were safe for the next few hours.

---

[30] Under Article 17 of the Hague Convention of 1907, neutrality restrictions, the 24-hour rule determined a belligerent warship could dock in a neutral port for up to 24 hours but could not receive provisions or coal except when it was in distress or did not have enough provisions on board to safely reach the nearest friendly port.

The medicine is soon taken away; in the meantime a watch officer went over to the Spanish cruiser “Cataluna” to arrange the usual formalities such as compliments, visits, etc., which are part of international courtesy. We had to inquire the senior naval commanders' names, town commander, mayor, etc., so the commander could report the presence of our boat and pay them a visit. With the help of French, communication succeeds, and when the officer on watch returns on board, our work is done and the crew can wash and bathe to celebrate the day. Large scrubbing tubs and are found on board "Roma", filled with water, and the scrubbing festival begins.

In the meantime it has gradually become day, the pilot must have reported our arrival, because now the Customs boat comes with officials, harbor police appear with their boats, and the nearby mole begins to fill with astonished people.

The soaping is over, the crew is beaming in their dress outfits; now the boat is being cleared up and made presentable. We had already started doing this at sea, but now everything is spic and span, as if we were to be inspected by a senior officer at home. Our boat number, which was specially painted for the purpose of our visit, is emblazoned in a giant size on the tower.

The sun has risen higher; it is almost ten o'clock German time, when the command is given "Ready to salute!" At the chime of eight o'clock Spanish time our German war flag rises on the flagstaff at the stern and the Spanish one in the maintop, simultaneously with it goes a twenty-one shots salute and then it is lowered again. Shot after shot rolls across the harbor as precisely as if we had been drilling for days and yet it was the first salute that our submarine had ever fired.

As previously agreed to an hour later the commander begins his visit. The immediate vicinity of the boat has been cordoned off by the police and military, because one never knows whether there are hostile agents planning attacks on us in a neutral country.

Our commander has hardly returned when the commander of the Spanish cruiser "Cataluña" comes on board to return the visit.

It was agreed with him our boat should have a convenient and safe berth alongside the "Cataluña"; directly under the Spanish guns, where one would hardly dare to attempt an attack.

His boat has hardly cast off when U35 follows his wake leaving its previous berth and making a wide arc through the harbor to tie up on “Cataluña’s” port side. The people on the pier think we want to put out to sea again, and rush to the pier head, but in an elegant arc we moor next to the "Cataluña", which is moored to a buoy.

Now begins comradely intercourse with the officers of this cruiser. First they come to us to look at the boat. Everything is shown to them and explained in more or less good French, mixed in with snippets of Spanish and English. After a tour of the boat, they were treated to port wine and biscuits in the mess hall, i.e. in our primitive bow compartment. There is not enough space for everyone to go down at the same time; so visits are made in groups.

Everyone is genuinely amazed at what they have seen, and each group always finishes drinking to the health of His Majesty the German Kaiser and his own: we consistently reply with the same politeness.

Then the gentlemen from the army appear whom the commander also kindly invited to inspect us during his visit to the port naval commander. They too are welcomed hospitably and everything is explained to them as far as possible.

By then it was noon — the commandant is still making visits. When he gets back into the steam launch, which the "Cataluña's" commander has kindly made available to us, his dagger falls out of its sheath and into the harbor as he steps on.

He quickly borrows another from one of our officers and departs. But a quarter of an hour later an equipped diving boat is lying next to us and after ten minutes the diver brings the lost dagger up. A small move - and yet proof of how much kindness and courtesy the Spaniards treated us.

For lunch we officers were invited by the "Cataluña's" officers to their mess. A typical southern meal consisting of countless small courses was set before us, with all sorts of wines; they even offered a German Rhine wine to accompany the main course in our honor.

An interpreter had also turned up in the person of the captain of an Austro-Hungarian merchant steamer lying there; that was very pleasant, because now, one could at least make oneself perfectly understood. We were happy to find even more sympathy for Germany and German character among the Spaniards.

Soon after the meal the officers' ladies appeared in the mess hall. Since conversation didn't go well, there was dancing, smoking, and drinking; visiting cards were exchanged, cap bands exchanged mainly for wonderful Havana cigarette tobacco.

Meanwhile, on board U35, the visits continued, one of the officers always had to be there to receive and guide.

The Consul himself came on board early in the morning and was greeted with the customary salute at flag parade. He undertook to send all telegrams to the embassy in Madrid, whereupon an extra train with the embassy staff was immediately dispatched and arrived in the evening. But I don't want to anticipate, I just want to tell the story as it unfolds.

Gradually the gatherings left the “Cataluña’s” mess hall and were back on board. Members of the German colony were just arriving and with them were the Consul and his wife. The joy of these Germans, they were some whom lived in Spain for years, to once again have German soil under their feet and to truly experience a bit of the Great War was touching. Everything was shown and explained to them and they were enthusiastic as probably seldom occurred in life.

Around the boat it swarmed all day with small boats that also wanted to see something of us. But the Spaniards had manned two cutters, in which patrols sat, who would not let any boat closer than two hundred yards without special permission.

When we said goodbye to those who came onboard, the Consul invited us to dinner in the most gracious manner, and since the commander had no objection, we two, an officer on watch and myself, agreed.

When it slowly got quiet on our boat, we took the opportunity to change our clothes. While we had walked around in dazzling white during the day, in the evening our home waters blue dress uniform was brought out, and proud as a Spaniard I almost said, we pulled ashore. Our constant companion was and remained our interpreter, the merchant captain. Under his leadership and accompanied by a few German gentlemen, we arrived ashore at about seven o'clock.

Southerners usually only appear outdoors in the cool evening hours and we found the promenade parade in full swing. We were surrounded as soon as we were on land and only with great effort could we gradually move through. People greeted us like old acquaintances, some crowded up to shake our hands and say "Vive el Kaiser" or "Viva Alemania". Three police guards who walked in front of us had to make a path through the crowd—and so we marched in triumph along the main street until the hustle and bustle got too much for us and we fled with the accompanying Germans to their favorite cafe.

It was here you first notice how thirsty you become after all the numerous spirits which had to be drunk out of politeness during the day; after a cup of black coffee, colossal quantities of cold soda water were gulped down. That was good! Here we were shown the first postcards of our boat, which were already being sold on the streets, and when we quietly expressed a wish, a whole lot of them were kindly dedicated to us.

But we had to go on, at eight o'clock we were invited to the Consul's, so back out into the hustle and bustle. Instead of the three policemen there were now six ready to pave the way for us; for the news of our presence on land had spread like wildfire, and a tremendous crowd accompanied us on our way. Amusingly, he led us past a cafe where the English and French consuls were sitting. I was sorry I didn't have my photographic equipment to hand, because I would have loved to capture their grumpy faces on the plate.

At last we arrived in front of our consul's house, which was decorated with a large German flag, and entered the house with a sigh of relief.

We were greeted happily by our hosts, but half of Cartagena stood in front of the house and waited for things to come. So something had to be done; people were not allowed to remain unsatisfied. The Consul quickly told us what kind things we could say to those waiting, and without further ado we stepped, proud as kings, to the parapet of the balcony and called down: "Viva España!"

“Viva Alemania!” came the answer, clapping hands, and with a grateful bow we avoided the looks and further ovations of the crowd.

A comfortable few hours followed—comfortable solitude, four of us facing each other, instead of the thousands who had crowded around us during the day; only German language, instead of the gibberish we had just crunched from the morning. It was a real relaxation, this quiet dinner, not the usual on our wobbling mess table, but tastefully and festively arranged and embellished with cold French sparkling wine.

It wouldn't have lasted long if we hadn't already been in the mood and due to the stimulating conversations about Spain and Germany, war and peace, time passed so quickly we were quite surprised when other German officers arrived at ten o'clock to pick us up because we had arranged with them to show us a bit more of Cartagena. We departed the friendly house with many thanks and greetings to our homeland and distant relatives.

Outside, meanwhile, the crowd had stood their ground the entire time and just as little as our police escort. They immediately accompanied us with the same touching enthusiasm as before. We had to say thank you more than when we strolling down Holstenstrasse at home, everyone said hello or attempted to say hello. We were getting tired—the day had been lengthy enough; so we sat down at a table in a coffee house on the street, where the crowd now filed past us. Our escort made sure to move them along. But in the long run there was no peace for us: as soon as we sat in front of the inevitable Negro, we were taken away again - two houses away was the bullfighters' club, and they took the opportunity to see us in among them as well.

Our usual interpreter, who seemed to be at home everywhere, introduced us with grandeur only possible in Spain, Prophets on the right, Prophets on the left, Child of the World[31] in the middle - our Capitan; who with an inimitable hand gesture bequeathed us to the gentlemen. We were forced into places of honor, naturally in such a way that those waiting outside could see us as much as possible through the large window—and we were entertained once more. Everyone drank to us, everyone had to be thanked, cigars, cigarettes were forced on us; whether you could smoke them or not, you had to put them in your pocket. A Spanish gentleman who spoke French was soon found, and conversation was carried on with and through him in all directions. Then each of us was solemnly presented with a Bullfighting Club membership card, which honor we received with due respect. The next thing was my shipmate was taken away from me: they wanted his autographs.

He sat in an adjoining room and wrote his name over and over and over again. Eventually it was too much for our interpreter, and he suddenly demanded five pesetas for signature for the Red Cross. The Red Cross made good business from his work.

---

[31] "Prophete rechts, Prophete links, das Weltkind in der Mitte" in the original. These are the final verses of Goethe's poem "Diner in Coblenz in the Summer of 1774," regarding the guest in the middle who is unmoved by the opinions of others around him.

The hands of the clock advanced onward. We needed to be on board at eleven o'clock. So I made a quick decision to get my writing companion away from his profitable work, with many thanks we said goodbye to our new friends and threw ourselves back into the crowd.

Military and civilian bands played religious music in the streets because it was the eve of the feast of Corpus Christi and there were a particularly large number of people present. The fresh evening air did us good, so on the promenade we stopped again in a coffee pavilion to calm down. Our table was quickly surrounded again; all the Spanish army acquaintances who were on board with us that day in uniform appeared and sat down with us, and again waves of southern enthusiasm ran high. One bearded captain who was particularly lively stands out in my memory; even his comrades called him the black captain.[32] Finally, despite everyone we had to leave. It could not be helped as I still had to charge the battery, make ready the boat for sea and diving. Everyone escorted us. Our interpreter's boat was waiting for us on the quay, but even here none of us wanted to part. When the boat pushed away, it was quite full, although actually only three people were allowed to sit in it apart from the boat seaman.

Thick and black, the crowd was massed on the bank, flooded by the glare of the arc lamps.

Not wanting to appear ungrateful after enthusiastic reception we had been given, we rose from our seats and shouted a resounding "Viva España!" across the harbor, which we followed with three German hurrahs. A never-ending clapping of hands, mixed with shouts of "Viva

---

[32] In Spanish the “Negron” moniker would be analogous to being considered a Black Sheep, a sentiment ranging from affection to critique.

Alemania!" was the answer. The darkness had long since hid us when the clapping of hands still resounded to us.

In the boat, the enthusiasm now swelled to the heights of the house. The black captain couldn't resist hugging and kissing us over and again. Everyone wanted to come on board; we had to deny them this wish; but in cordial friendship we parted.

Then a few brief orders: "Get the watch up for charging!" And above all, just one thought: Sleep! Good Lord, we were tired.

But oh, sleep? The embassy gentlemen arrived on board and are filling our compartments, so there is no free space. A brief introduction, an apology for the disturbance, and back on deck. But alas, there is even a lady on deck, and the other officer on watch who was not on land catches me and introduces me. So in God's name I pull my tired bones and heavy head together, entertained – made conversation - as best as I could. After a while I hear the voice of the lady — it was the wife of the German attaché — say to me as if from afar:

"Are you very tired?"

"Oh no, on the contrary, Madam. For what reason do you ask?"

"Well, you have your eyes closed while you talk!"

I have to laugh: "Excuse me, but Madam is correct, and I can only very much beg your apologies and permission to retire."

Into the hatch and throw the dress coat in a corner, “Wake me up if something's going on!" and I fell asleep.

I thought I had slept for five minutes when someone shook me awake.

"Commander's orders: prepare lower deck ready for diving."

I stagger to my feet — a moment's reflection, the order will be carried out. An old service jacket somehow found its way on me, and I go to the commander, who is standing on the tower, and tell him the boat is ready for sea and diving. A fantastic picture opens up: Outside, on the other side of the mole, only a few miles out to sea, it is bright as daylight, searchlight after searchlight wanders across the water. This is the pack waiting for us outside and has flocked from all sides to catch us. Now it's time to get through it.

The boat slowly detaches itself from the "Cataluña", on board all the lights have been extinguished, it is three o'clock in the morning according to our time; friendly Spain accommodated us for twenty-two hours.

U35 silently glides out of the harbor. The enemy out there can't know when we're coming and the Spanish torpedo boats are standing in front of the harbor, making sure we're not interfered with in Spanish territorial waters.

But we won't let it come to that. Hardly are we out of the harbor and in deep water, still shrouded in dense darkness, than we evade the searching searchlights by diving and head for the open sea at a safe depth. By three o'clock the next afternoon there is no enemy in sight. No one saw us but we saw quite a lot.

How much the enemy was in the dark about our whereabouts is shown when a French auxiliary cruiser entered the port of Cartagena again the next afternoon, searched it without anchoring and immediately put out again.

A few days later, however, the German newspapers reported:

Berlin, July 5th. (WTB) SM submarine U35, which brought to Cartagena a handwritten letter from SM the Emperor to SM the King of Spain and medicines for Germans interned in Spain, has returned after successfully completing its mission.

On this journey, the boat sank the armed French steamer "Herault" and captured a gun.

Chief of the Admiralty of the Navy.

Alarm Stations in U35

The Last Shot.

Course homeward! We must be there in a few days: we have no more torpedoes, no more explosive shells and only one shell on board, apart from a few saluting shells left over from our visit to Cartagena.[33] So we can no longer do anything at sea and avoid everything that comes our way as far as possible or, if that is not possible, we hide underwater.

It's a Thursday. The sea is as smooth as a duck pond and we intend to get a good deal closer to home harbor today.

At nine o'clock a steamer—two masts, one funnel—is reported to the commander.

"Steer out of his way!" is the reply.

So we alter our course so that the steamer's course and ours form an acute angle.

We let him pass us – and the funnel and mast slowly disappear under the horizon.

When they are no longer to be seen about half an hour later and this is reported to the commander, he orders: "Return to the old course!" Accordingly, the course is changed again and we hope to remain undisturbed for a while. But after an hour the steamer is in sight again.

"Apparently he doesn't want to escape his fate," says the commander and orders "keep at it!"

---

[33] U35 had already sunk 18 steamships, 21 sailing vessel; 26 of which were Italian flagged vessels of which no claim could be made by Germany not being at war with Italy at this time.

We turn towards him, but soon he notices us, turns and runs away.

"Both oil engines full speed ahead!"

The hunt begins. At first it seems as if we are not getting any closer, but after half an hour we have the certainty that we are slowly but surely coming up. If nothing else comes into view and he doesn't shoot, then there's still hope of catching him.

After an hour we are about five thousand meters closer.

"Gun crew on deck, bring a saluting round!"

The crew appears on deck laughing and puts the cartridge in the breech.

VROOM, the gun salute booms over the water. The steamer feels honored and continues on.

Since he doesn't react, there's nothing left but to keep chasing him. The distance is getting shorter — involuntarily everyone is thinking of the shell we still have. Slowly we got a thousand meters closer!

"Load the shell! Fire over him with highest elevation!" orders the commander.

Said and done! Our last shell lands in the water far in front of him.

Everyone stares intently through their glasses. It was our last shot. What if he doesn't luff on that either?

Suddenly everyone's face lights up—a joyful shout goes out: "He's blowing off, he's stopping!"

And really, a white cloud of steam rises vertically up the funnel! The last shell worked.

"Both engines ahead half! Ten degrees to port."

Now the English flag is flying at the flag pole. And that too!

"Signal: Abandon ship!"

It flutters merrily in the wind — two boats are launched into the water on other side.

Now it's time to be careful so that he doesn't deceive us after all. The distance must be maintained, perhaps he still has guns in disguise, because an Englishman usually doesn't stop after one shot.

Now the two boats set off and crawl across the water like snails.

"Where is the captain?" we call to the next boat; of course he is sitting in the other one.

We stop and wave him to come alongside.

This happens, and soon the captain stands in front of us and hands us his ship and cargo papers.

They are quickly looked through: It is the English steamer "Teano" from Hull with coal to Naples.[34]

It is good. The captain can go back into his boat and is dismissed.

"I hope you are content with me!" he calls out to us as we set off.

Hello! What is that? "Hold the boat, we'll take you with us!" the commander suddenly orders. "A man with a pistol deck! Stand next to the boat!"

[34] The 1913 built Teano, 1,907 tons. There were no casualties.

Cautiously, offering the smallest target area, we approach the steamer slowly, very slowly, constantly scanning it with the glasses. Shouldn't it be a submarine trap? The captain's statement seems suspicious to us: so he must accompany us as a hostage. But nothing stirs on the steamer. Down in our boat everyone is at diving stations, the oil engines are turned off, we drive electrically to get away quickly just in case. We're already a thousand yards away and there's nothing to see over there. When there is no life beyond three hundred meters, the boat is thrown off.

"That is only for my security in case of Baralong!"[35] the commander calls out to the steamer captain, who makes a long face and is happy to finally be released. Only now does he seem to realize his situation and our behavior.

"The commandant asks," reports a stoker.

"Please go over there with your men and open the sea cocks!"

"Yes sir — it will be done!"

Quickly down again, a few able NCOs and stokers are called—we don't need to take any tools with us, they can be found over there—and then into the thing!

Apart from the boat seaman and I, only a non-commissioned officer and a stoker are on board, the small boat cannot carry more. A second non-commissioned officer and a rating are to come over on a second trip.

But oh dear! There is no sea ladder, and so it is necessary to climb up the hanging boat ropes to the high boat deck and from there try to catch the gently rolling steamer while swinging. It works, I'm the first to get up and help my

[35] The well known 1915 incident in which the Q.ship Baralong sank a German submarine and shot the escaping submariners.

people across. The dinghy rows away again and fetches the other two.

Access to the engine and the boiler room is easy to find, for this not the first steamer to lose its life in this manner, rather the eighth ship that I have sunk in this way. They are all built according to scheme F. With the help of an electric flashlight, the sea cocks were soon found. The exterior valve is closed and the interior valve cover is removed. Once this is done, the exterior valve is opened again and the water floods into the engine room in a high arc. It is unpleasant when there is only one exterior valve connection, then you have to press it tightly with a rod, a poker, or whatever object you can find until all the screws are loose and the man is out of the bilge. If you then let go, the valve, lifted by the water pressure, flies in a high arc into the compartment accompanied with the rushing water.

This steamer has two sea cocks, so the work was relatively easy.

We're still busy with it when my other two men, whom the dinghy brought over on the second trip, appear in the engine room. I have one of them remove the bilge lines valves so water that floods in can spread over all the compartments. Without that, only the engine and boiler room would fill up and the ship might float due to buoyancy from the cargo holds.

I myself go into the boiler room with the other man and open all the bunker doors to clear the way for the water. As soon as we're done with that and back in the engine room, I receive a message: The second valve is open. Now the first valve is open and water shoots into the room in a mighty jet four to five meters high. You get a little wet, but that's irrelevant. Then I hurry on deck and wave to the boat, the commander then knows that the sea cocks are out and the water is flooding in.

Now below again. A man is placed at each bilge valve box; valve after valve is pulled out and thrown somewhere. The pantry offers plenty of tools and lighting. If this key doesn't fit, then that one does, and the work goes quickly.

I walk on and turn my interest to the propeller shaft tunnel. Sometimes there are manholes or doors leading to other compartments that have to be opened to let in water. I show the fireman what he has to do: "The manhole cover is removed and the three valves that are in this line have to come out."

Then I return to the engine room, the first valve box is ready. I take the non-commissioned officer with me to the boiler room, because I have another one there too. A bilge valve box is spotted half hidden under coals. That too is included. To be on the safe side, I also check the fires and the water level in the boilers. But everything is in order, there is no danger that a boiler will burn out and blow up in our faces.

The water in the bilge rises and the ship begins to list. Already water slowly begins to cover the floor plates. But my people continue to work calmly standing in water. It shoots gurgling through the valves into the piping and spreads to all compartments; as a result, it no longer rises so quickly in the engine room, and the work goes ahead briskly. Only the man in the shaft tunnel has to give it up, because the tunnel is deeper there than any other compartment and he's already up to his waist in water. But two valves have been removed and the manhole cover is wide open.

The work down below is done in thirty-five minutes. Now it's time for the drains. Anything that doesn't want to open is smashed without further ado, and soon the water wells up through these openings. They are roughly level with the waterline and only really get going when the steamer begins to list even more. And he will soon.

We are done and climb back on deck dripping and pitch black with coal dust and soot. We wave our dinghy over there and climb into it, but it's not quite as distant as before because the ship is already down considerably.

On board I report what happened to the commander and assure him the steamer is certain to sink. It takes time, about five to six hours, but no one can bring him home because the engine room and boiler room are full and all the draining equipment has been destroyed. Even if help comes and tries to tow the steamer away, it will sink away from under their hands.

In the meantime, a watch officer has inspected the captain's quarters and taken all papers that seemed valuable for inspection.

After an hour and a half we continue our journey home satisfied, in the knowledge that we have placed our last shell well.

If Englishman had only known the shot was our last shell.

Would he have stopped so quickly?

Alarm Stations in U35

How we got the small French cruiser "Rigel"[36]

We were standing not far from the Algerian coast, and at dawn we saw a plume of smoke on our heading. After an hour we were still not close enough to see the steamer from which these puffs of smoke had come and we gave up the enterprise and head for a sailing ship whose masts came into view on the horizon.

With the little wind that prevailed, we soon reached it. It was the American five-masted gaff schooner "Van Allen Bughton" from Boston, empty from Genoa to Cadiz. We had to let her go because she was empty and neutral, we then turned back to our old course.

After an hour and a half, two typical warship masts come into view. We spot a small cruiser of the "Arabis" class as the funnels come out and we immediately dive.[37]

The commander decides to attack him and advances to attack. Forty minutes after the dive we find him within shooting range and the torpedo leaves the tube.

After forty seconds a detonation takes place and everyone's eyes shine. Another warship. Trade warfare is good and has more impact on the outcome of the war, but we're happier when we can kill a warship. It is probably the feeling of dealing with an equal opponent, having to fight with the enemy's military power and not with his financial strength.

---

[36] Events in the chapter occurred on Oct. 2, 1916.

[37] The two stack, high freeboard Arabis class do appear as a prewar colonial 3rd class cruiser or gunboat. The class tonnage of 1250tons, modest dims and previous experience with similar vessels indicates Fechter is magnifying the target for the reader.

Everyone is still curious to see if he will sink. The hit sits in his port side, and initially he has a strong list there. But a few minutes later he straightens up again and then lists over to starboard.

The ships of this class must have an impeccable bulkhead layout, because they will not sink if they are hit by a torpedo, as we had already noticed when sinking the "Primula".[38] And this one has its hole on the port side, falls over to starboard and continues calmly on a northerly course.[39]

It's impossible to get to him again. So what to do? The commander calculates as follows: he won't get as far as the French coast on this course; our torpedoes are too good for that. He can call at Port Mahon, but there is little prospect of being able to repair it there. The nearest major port is Algiers, so that's where he'll be returning; only heading north to try and evade our sights. So we will take a course east.

As it is shortly before lunchtime, the lunchtime meal of peas and bacon is eaten in peace and quiet. After the pros and cons have been properly repeatedly discussed, the commander looks out again at two o'clock. And right: his calculation was correct. Coming from the north, the cruiser now steers south and thus runs directly in front of our bow for the second time.

Two guard vessels, trawlers, are with him now. So he has summoned support after all to escort his voyage and stay close by to provide any help that may be needed.

---

[38] HMS Primula was another Arabis class sloop attacked and sunk after a strenuous resistance on March 1, 1916.

[39] Rigel's commander, Capitaine de Frégate Pierre Jules Marie Vandier reported the torpedo hit the forward coal bunker and created a hole 4 meters wide. The bulkhead immediately adjacent held to the flooded compartment held.

We go on the attack again and have to close slowly so as not to get to alert him.

The maneuver was successful: at 3:15 the second torpedo left its tube, and a few seconds later there was a crash. -— Hit again, now he's going under for certain.[40]

"Got to meters," calls the commander. "Trim forward."

VROOM, Bang, Bang - there are crashes and bangs in our immediate vicinity, and the boat wobbles as if someone wants to shake it off.

"Watch valves and bilges!" I give my order to all compartments.

Bang, bang, Fum! There's another crack next to us in the water.

"Water ingress through the forward trim tank," comes the message from the bow compartment.

"Trim aft! Water in the bow compartment — I'm going forward," I call to the commander up to the tower and jump forward.

The front compartment is shrouded in deep darkness, only at the very front someone is just switching on an emergency lamp. By their light I spot a dripping figure, the Senior W.T. operator, who has his diving station here.

"What's up?"

---

[40] Capitaine Vandier reported this torpedo ran on the surface where one could clearly see its bronze cone and the circle of red paint painted on its air tank. The detonation tore off Rigel's stern. Damage control nearly saved the vessel which sank 9 hours after the second attack. Vandier believed the vessel could have been saved if the slow leak in both an auxiliary watertight door and boiler room drain were stopped. 17 men were killed in this action.

"When there's a bang outside, water comes into the boat through the trim tank sounding tube and I sat on it."

"That's all?" I call out, laughing. "Then tighten the screw connection from the sounding tube! Tighter so next time it won't fly in your face when you trim forward!"

I'm already back and describe the completely harmless incident to the commander.

“The front trim tank was full, and when trimming forwards, the water pressure blew up the loosely mounted dipstick fitting. The water got to the electrical wiring, so the light went out, coincidentally there was a bang outside at the same time. The Senior W.T. operator W. bravely sat down on the sounding tube and closed the hole, of course he's soaked."

Although there's another bang outside - now a little further away than before - the commander has to laugh heartily.

We don't even think about showing up. We change our course underwater and run away from the enemy as fast as we can with the empty battery.

Finally at seven o'clock, after we had shaken off a few destroyers, we can surface.

Cool and fresh, the fan blows pure pleasure into the boat, and everyone is happy that things went well again.

"Well, how are you?" I ask the Senior W.T.-operator, who has already changed.

"Oh, Chief Engineer," he says innocently, "the fresh air is better now than it was five minutes before a hero's death!"

Fips the monkey.

We had forced a steamer to stop; its crew sat in the boats and sailed away. Our boat was surface alongside him at some distance. The deck gun was being manned to give it the coup de grace and send it down. The off-watch had come on deck to watch the spectacle.

"Cannon ready to fire!" – "Aim for the waterline!" – "Open Permission!"

The first shell hits the steamer's waterline with a crash, and it is followed by several well-placed hits at short intervals. Water gurgled through holes in the holds and into the engine room.

"Cease Fire!" "That's enough, I think," says the commander with a smile.

"How long are you giving him?" he turns to me.

I look at the steamer and then at the holes: "A good half hour!"

"Then we don't need to stick around any longer!"

"Starboard oil engine tie-in ready to move!" he orders, pleased with the job done.

"There's a Monki over there," the voice of an off-watch sounds up to us from the deck.

All eyes turn to the sinking ship.

That's right, up there on the set dinner table, from which we so rudely startled the captain and his officers, is a little monkey sitting nibbling and eating.

"Sir, We must have it! He mustn't drown!" begged the men.

A moment's silence.

"Stop the starboard oil engine! Put out the dinghy!" It was done quickly, and a watch officer and two men hurriedly pull over. They soon arrived, quickly up the sea ladder and over to the monkey.

However, he does not greet his rescuers as enthusiastically as one should expect, but is snappy and malicious. But there is time to lose: with a short jerk the hook of his chain is torn out of the wall and the good Monki is dragged backwards rather than forwards across the deck.

The main difficulty comes getting him down the gangway. He comes willingly to the last rung of the rope ladder, but then he doesn't want to part with his ship. He holds on to the rope with all his strength, suddenly bites the fingers of his new owner, who is holding him by the chain, and with a bold leap he is back on his ship. He performs this maneuver once or twice until the dinghy crew gets bored. Without further ado someone grabs his neck and holds him outboard into the sea. That works; now he sits very still and lets himself be pulled over to us.

General laughter greets him; he looks angry at anyone who comes near him and bares his teeth,.

With great difficulty they get him through the hatch, and now he's sitting wet and freezing in the control room, where we initially tied him up.

Attempts to dry him or take him to a warmer place is out of the question: any bold person who dares to do this will leave with bitten and scratched hands.

He crawls into the furthest corner and gives a fright when the oil engines starts to bang into operation and the blower starts screeching—as soon as some new, unfamiliar noise commences. He withdraws timidly into the darkest corner, from which he curiously peers out once he has slowly gotten used to the new noise.

He even refuses food on the first evening.

But the next day he is already more docile. Every now and then this or that man arrives with a piece of bread, a potato, even nuts become visible. Fips quickly snatches the food from him, but then immediately prepares to attack, so that everyone prefers to keep their hands intact than try to win their friendship.

Nobody knows how he got his Wilhelm-Busch-name[41], but from the moment he appeared on board with us, his name was Fips. I think the commander called him that when he didn't want to go through the turret hatch.[42]

In the evening he is taken to a quieter place in the stern compartment. The torpedo launch rails run below deck, where he can refresh his primeval memories, and he eagerly jumps from one to the other. Some people hook their night hammock under these rails. They try to play with Fips a little more. But he doesn't react to any advances, and disappointed, the men lie down.

After half an hour they are sound asleep and snoring with their mouths open so that it can be heard almost in the engine room.

Fips is also infected by the general tiredness. He jumps sleepily from one rail to the other. Finally he thinks he has found the right spot. He crouches down. Then it becomes

---

[41] Most likely drawing on the mischievous antics of characters drawn by German humorist, poet, illustrator, and painter Wilhelm Busch.
[42] Fips in German is often utilized as a nickname originating from attributes describing personal characteristics.

clear to him that he still feels a human emotion - and without changing his position, he lets his feelings run free.

In the chorus of snoring, a voice suddenly stops. One hears spitting and wild cursing, mingled with the screams of the monkey, which is getting its beating because it remained true to its jungle habits and had not yet gotten used to the fact that there are people on a submarine who are under it with their mouths open sleep.

After a few days, Fips got used to our life on board. He's free to roam now; under the crew he has found his friends, who are even allowed to pet him if they give him something to eat. He also accepts food from others and then snatches it away from and attacks them.

Wherever he finds something edible, he steals it. Nothing is safe from him. Flour, raw potatoes, sugar, eggs - he steals everything and then takes off with his loot in long leaps.

In doing so, it is not at all important for him to seek his way over the exposed live current switches or up between the levers of the running engine when escaping; he's over it with truly monkey-like speed without taking the slightest damage.

Once he even managed to escape through the whole boat up to the tower, holding a raw egg which he had stolen from the galley, without it breaking. It was only here that he artfully opened it and drank it down.

He only escaped up the tower and on deck when the weather was very fine. Then he was up most of the day, warming himself in the sun.

At first we were afraid that if we had to dive quickly he would stay up and drown miserably, and one fine day it really seemed so far. We had to dive and Fips was still out; we tried to catch him but he got away from everyone so we had to give him up. The air hissed from the vents, the boat went lower and lower. There—a leap, and Fips was in the tower; it had become too uncomfortable for him. From that point on, we didn't bother to get him into the boat before diving: as soon as the vent spluttered open, Fips was the last to jump into the boat by himself.

He did the same when we sank steamers with artillery fire. At first he watched it all curiously. But as soon as the cannon discharge, he was inside; he couldn't stand the blast noise.

At first it was a nuisance to find a place for him to sleep. Boxes were placed everywhere, softly padded, in the quietest, driest corners, but he wouldn't get in.

Every day he sat in a different corner, now here, now there. One fine day or rather night I wake up and I feel something warm, soft, and hairy in my arm. It was Fips, who simply packed himself into my alarm bunk.

From that day on he was my constant guest.

As soon as I lay there wanting to sleep, be it day or night, Fips appeared, laid his head on my shoulder, stretched out by my side and slept calmly and peacefully in my arm beside me.

Despite some difficulties, the presence of Fips brought with it a very educational moment: nothing was allowed to be left behind, neither small tools, notebooks, pencils or anything else. He took everything; what he couldn't eat he carried away never to see again. Once he ate an ink pen complete with skin and hair[43]; his snout was blue from it for days. He liked to pour out the ink bottle over official diaries as soon as one turned one's back. He had no respect even for my measuring instruments. One day he stole my glass aerometer. It was taken from him - and put back on the spot after he received a beating. He jumped away and calmed down, reassured the petty officer went back to his work. There's a clink behind his back, and the only aerometer I had on board lies in shards on the floorboards.

Fips also had character. If he ate something and got hit by someone, he didn't attack his tutor afterwards, but those who had stood by while he was being disciplined and laughed at him while he was being beaten.

He liked being on deck best. He jumped all the way to the bow and then through hatches and holes into the empty spaces. He watched the life and activity on deck attentively and tensely. Now he stuck his head through this, now through that hole, looked around and dived away again. On the whole he was really amiable—even if he couldn't entirely deny his ape nature. One evening a few tired swallows settled on our boat deck. As soon as they were sitting, Fips was there and jumped up with one in his mouth and into the empty space in the bow where no one could follow him. He received a miserable beating for this outrage, but he couldn't go against his nature.

For a while we had made him accustomed to appearing in the bow compartment with us officers every midday after the meal and picking up his share of the meal in the form of fruit, biscuits, compote and the like. At first it went quite well. But one day we got busy and had to give up eating. Fips appears, looks around, finds nothing and therefore

---

43 “Mit Haut und Haaren” in the original. German colloquialism meaning “complete” or “completely through”.

attacks the commander's desk. The result was terrible. Pencils eaten, ink spilled, service letters torn up, cigar holders broken—everything topsy-turvy. The result was that we had to break the habit of visiting him at noon. But it took a long time, because he liked it too much, and many a time we had to chase him out of the front compartment.

He became very good friends with some of the crew. When the weather was nice, one of the front hatches was always open. Fips sat on the edge of the hatch and waited until you stuck your head out of the hatch. He would sit on your neck and bite you, very friendly: he wanted to play and was happy if you shook him and teased him.

Fips's cleanliness was also taken care of. He was often combed with an old, discarded comb, and when that couldn't keep up with all the dirt in which hung about, he was even bathed and soaped, but he had no sympathy for that; he withdrew from this event and often fled into the furthest and deepest corners where no one could follow him.

So Fips, with all his habits and bad manners, had become a pleasant companion and good friend on two long journeys, who helped us through many a bad mood: you had to laugh in the deepest impassivity when you saw how funny and fast all his movements were how cleverly he performed every prank.

But autumn came. Clattering, freezing and sometimes coughing, Fips sat sadly in a corner and was only half as funny as in the warm summer. We didn't want to let him perish in misery; after all we had grown too fond of him for that. So it was decided to donate it to the Zoological Garden in Berlin, and today you can find a cage there with a sign: "Gift from the crew of U35".

But I think Fips liked it better with us in the boat.

Alarm Stations in U35

With both engines broken down.[44]

We were in the English Channel and engaged in cruiser warfare. On the way there we had happily survived all the dangers the Englishman created for us. In sight of Dover and Calais, his brilliantly lit guard even allowed us to surface and load for three hours. The first day under the English coast also brought us two steamers. So we were in good spirits doing our work.

In the evening of a thick and foggy day I charge my battery with one oil engine, because on such an unsightly night we prefer to spend the hours of darkness underwater.

Right now I'm standing in the engine room and watching the current going into the battery. The ammeter slowly decreases. A look at the revolution indicator. The turns are still correct. Why is the current decreasing?! Everything is carefully checked, thermometer, manometer — but no error can be found. Then it comes down to touch and tenderly stroking the hand glides from cylinder to cylinder.

Ow! Cursed! Fingers twitch back, burned. A piston cylinder is hot. A Sight! — A few flicks and the switches spark, the engine stops! I go to the commander, tell him how things are, and ask to be allowed to finish the loading with the other engine. He agrees, gives up on traveling that day and leaves it to me to finish charging the battery with the other engine.

---

[44] Events in this chapter refer to activities in 1915 prior to entering the Mediterranean. Voyage of June 1915 being the most probable timeframe.

Soon it is turning at a regular pace and the current is going into the battery again. Charging is finished when the sun sets, thick fog lies over the water, but now we can safely spend the night underwater, because there is nothing more to be gained being up there. At most we can be surprised or even rammed in the fog. So we dive and can now devote ourselves entirely to our port engine without any disturbance.

But the damage is worse than I expected. The piston seized in the cylinder. The areas are smoothed out as best as possible and the piston itself is carefully examined for cracks. There are none - fortunately for us; otherwise it would no longer be usable and we would have to deactivate it. The work of straightening takes the whole night. When we surface at four o'clock in the morning to air out the boat, I turn the oil engine with the main electric motors so the running surfaces can smooth each other out again with a generous supply of oil.

The twenty minutes that I need for airing are soon up and the waters have closed over us again, for although it is no longer foggy up there, it is still pitch black and not very visible. The piston is inspected again and every piece is examined in detail. — But what is that? Previously it had turned blue and was a bit rough; now there's a crack about two millimeters wide across its skirt; the internal stresses had burst it while turning. So it doesn't help, it has to be deactivated. Get to work!

The middle watch who has been up since midnight begins their work a little tired and sleepy. Valves and bearings are removed, pulleys attached and everything prepared to restrain the piston in the cylinder so that the rotating crank no longer touches it. The question is how? Special devices are not available for this purpose, a makeshift must be created. Of all the material available flat iron and round iron is selected and tested for durability and strength. Finally we found a piece that seemed to meet the requirements. In my mind, I can already see this emergency construction completed. Actually I should have a forge fire, but here the iron has to coldly submit to my will. A bar is placed over the cylinder combustion chamber; it has to support the weight of the piston with strong screws. To ensure the piston cannot fall out if these screws loosen due to engine vibrations a second bar must be placed under the foot of the connecting rod and also fastened in some manner, if possible with existing screws. A quick estimate — an assessment. Result: it will last.

We surface at six o'clock in the morning. The cool sea air blows freshly through the boat. It is clear topside and the sun promises to drive away the last faint layer of haze that still hangs over the water. The first thing to do is charge the battery, because its content has dropped to a third due to our nightly underwater trip. So we start the starboard oil engine for loading. This happens — and she begins her work with calm, even gait. Soon it will be warm enough to develop its full performance. And this is already being increased.

Rum — boom! It suddenly announces and at that moment the whole engine room is a smoke and fumes. Out with the switches. The starboard engine also seizes.

What now? First let's see what's going on. The spot is soon found: a connecting rod bearing pin which actuates the air pump piston has snapped, the bearing has broken open and the rod is bent.

Reserve? There isn't one!

That's cheerful. Both engines out of commission, the battery a third full and in the middle of the steamer route in the English Channel, where the guards are quite good according to previous experience.

First off to the commander and report our bad luck to him.

He wants to proceed out of the steamer path with the least amount of electricity, so that we are disturbed as little as possible during our work.

Because we can no longer stay underwater all day, our electricity supply is not sufficient for that. And what to do if a patrol now attaches itself to us and forces us to do it? But why paint the devil on the wall.[45] Taking action is more appropriate. All superfluous light, all unnecessary auxiliary motors are switched off. Anything that needs to be operated manually is disconnected from electricity. It looks dark and sad in the boat. Only one lamp burns in each compartment. But there is light in the engine room and everyone is at work. The off-duty watch, which has only just been relieved and has hardly had breakfast, has to come as well. Even seamen who have the off-watch are called upon to lend a helping hand. The more workers, the faster it goes and every minute is precious.

On the port side we already know what we have to do. The starboard side crew are employed and distributed so everything goes hand in hand. The sad remains of the rod are cut off with a saw; because removing the slider would take too much time. The slide itself is taken hold of. It's a good thing the engine has two identical pumps that push into a common channel, so one can do the work for the entire engine. Everywhere air could escape, sheet metal should be placed to seal the channels from the outside. But where do you get the sheet metal from? There is none. Oh what, none available! The engine floor plates are just about

[45] “den Teufel an die Wand malen” in the original. German colloquialism for “Fearing the worst.”

suitable. Out with those things. Hammer and chisel give them the right size. At the back, two men are constantly drilling hole after hole through them with a hand drill because they have to be screwed into place. To me they seem too weak to withstand the pressure that is supposed to be on them. A few angle irons were quickly cut out of the floor plate supports and lashed over their former plates.

My people were a little uncomfortable with the situation, and there was a certain concern in their eyes. Hey, I think you don't want to get discouraged? The situation is saved with a few lazy jokes and they can laugh again. In the middle of it all, throwing in a catchphrase here and there, helping here, arranging things there, cheerfully reprimanding someone who is clumsy, I soon get them back to their self-confidence and the work just flies.

From time to time I ask the tower watch if there's anything in sight. But nothing more than clouds of smoke have been seen so far.

How time flies! I thought it had been a good hour or so since the starboard engine failed, and now it's ten o'clock in the morning. But we've also done a lot.

The floor plates now fit and are screwed tight, the slider holds tight, and an hour later I can report to the commander: "Starboard is being engaged for loading!"

And really, it works. Not only does it spin, it loads, if only half as much as usual, but it loads; we can generate electricity for underwater travel again!

Now the off watch can step away and indulge in a well-deserved rest. They have been at work since midnight, except for a short breakfast break, and they have done a splendid job. They should not take over at twelve noon as usual, but only at three o'clock, because shared pain is pain halved, the watch stays on duty for three hours longer from six to twelve o'clock.

We're almost there on the port side too. The piston is stuck. There are only a few more guides need to be sealed, and then it will be clear again. The seamen, who have behaved well in the unfamiliar environment are sent away to eat. At twelve thirty I go to the commander and can report to him that we've finished it.

Everyone breathes a sigh of relief. In a few hours we can have a full battery again and then the Englishman can slide down our hump.[46] We turn around and approach the steamer route we had previously fearfully avoided, as both engines can now run at half speed again.

But now ambition comes: this thing annoys me. What is a half speed boat? Nothing half-hearted, nothing whole![47] The port side is probably doomed to remain troublesome on this voyage. But the starboard side? Something has to be done with that.

I wander through the boat searching, pondering, pondering – finding. That's the way it has to be! – My non-commissioned officers look at me with a shake of their heads as I explain to them this thick eyebolt can be reworked into the upper shell of a new connecting rod, that this thick-walled piece of brass tubing, the inner diameter of which coincidentally matches the diameter of the sawed-off connecting rod socket, will make the rod, and a small bearing, taken from another engine and cut in half to half its width will become the lower shell. They don't think it will last. But they set to work and soon grew fond of the

---

[46] "den Buckel herunterrutschen" in the original. German colloquialism meaning "Go jump in a lake."

[47] "Nichts Halbes, nichts Ganzes!" in the original. German colloquialism meaning "Neither here nor there."

idea. After a day and a half I have my new connecting rod in my hands. Now it's time to attach it. The next night we are underwater. This opportunity is used to assemble this strange fabrication. When we surface in the morning, the oil engine is resplendent in all its glory, and now the big moment has come: trying it out. It runs perfectly at half speed, only higher revs don't seem to please the new machine part. Therefore as a precaution we only stay at these speed levels for a short duration and then resume a deliberate half speed.

The following night our connecting rod is removed again, the hollow tube receives a core of exact length and strength, the attachment to the old connecting rod stub is strengthened, and the next day the starboard side runs again without any problems at all speed levels and also holds for the whole thing voyage. Today, I believe, this unique drive rod is kept as on display in one of our imperial shipyards.

Alarm Stations in U35

Final Word.

I could say a great deal lot regarding my experiences. Because almost every day on U35; brought impressions that would be worth dealing with in detail. Little incidents, small things in dealings with seniors and subordinates, great and vivid events have remained in the memory. But all these details would ultimately pass by the reader without a trace because of their apparent similarity. What has value for the participant is often not of interest to the outsider. That is why I have tried to assemble together here the most characteristic moments of my work as an engineer, in order to give a brief outline of the life and experience of someone who represents his world at the innermost core of this small submarine community. If I had contributed such a thing to convey ideas of what goes on a submarine to accomplish the feats that fill the world today, I would be satisfied and hopefully my readers would be too.

Alarm Stations in U35

"U35" on the Hunt[48]

Lothar v. Arnauld de la Perière

Before the war, I was a watch officer on the "Emden", and at the very beginning of the war I was adjutant to the Chief of the Admiralty Staff, Admiral von Pohl, in a position that is indeed honorable but in no way satisfactory to me. I wanted to get out of there and soon, otherwise it might be too late. It wasn't until the spring of 1915 that I managed to get into submarine arm. Training as a submarine commander lasted until autumn, and finally at the end of 1915 a front-line submarine became available for me, namely U35 in the Mediterranean. I took over U35 from Korvettenkapitän Kophamel, who was intended to lead the Mediterranean flotilla.

There was someone else with us who wasn't having a hard time at headquarters either. August Haiungs from Dithmarschen. Tall, lean and sinewy. He didn't say many words. But you knew what you had in him. He tormented me to take him with me. He became my orderly and later specialized in boarding ships. Like no other, he climbed on board to determine the charges and attach the explosive cartridges. He accompanied me until the end of the war.

My first trip to the front began at the beginning of January 1916, at a time when unrestricted submarine warfare had already been stopped again because of American protests. Merchant ships were no longer allowed to be sunk without prior warning.

---

[48] Published by Bertelsmann-Verlag as part of their "Spannende Geschichten" or Exciting Stories series. Short length paperbacks focusing on military endeavours from recent wars.

U35 on the Hunt

The maritime area between Malta and Crete was assigned to me as the area of operations. "You don't need to sink anything; the main thing is you bring the boat back from your first patrol in one piece." With these words our flotilla commander in Cattaro dismissed me. –

On January 17, 1916, I stopped my first steamer, the English "Sutherland," and sank her by gunfire. Everything went smoothly and now the spell was broken. The next fellow to come into view was an armed Englishman who, in response to our warning shot, turned and took up the fight. During the pursuit another steamer, flying the Dutch flag, appeared and had to cut our course. So I had no choice but to shoot him across the bow and let the other one go for now. At the signal "Send boat" a boat was launched. There were 4,000 meters between us and the steamer; it would be dark before the boat reached midway. So I had to proceed towards it. Remembering all the good advice of my comrades, I did this very carefully, diving with my periscope extended wide. We circled the steamer a hundred meters away. Nothing moved on board with him. The name "Melanie" was written on the bow. The crew was in the ship's two boats, which were, as usual, 800 meters aft. We looked at the steamer very carefully, I took turns looking through one periscope and my watch officer, first Oberleutnant zur See Steinhauer and then my Chief Helmsman Neumann, both of whom had already completed many trips. We agreed: it was harmless, so we can close in for an examination. I'm about to give the necessary orders when I instinctively feel inhibited and my mistrust returns. I will certainly surface, but not here, 100 meters abeam of the steamer, but astern by the two ship's boats. Once I have the two alongside, he can't shoot without endangering his own people. So I went there and gave the order to surface and man the cannon. We were already on the surface, still 30 meters away from the boats, the gun crew was just about to man the gun when a hellish fire broke out from the steamer. Flaps fell, the first shells whizzed hard over our heads, others whipped the water onto our deck, interspersed with raging machine gun fire.

“Flood,” “Quick Dive,” “Gunners Inside.” Behind me, the brave gunners rumbled through the tower hatch and one of them threw it shut. I was about to go deeper when one of the people shouted: "Leutnant Launburg is not down here!" I couldn't let our watch officer drown, so I went up again, opened the turret hatch and he came down accompanied with a good shot of water. Now let's go! We immediately dropped down to 50 meters. It was clear again. Why? The only explanation is that the Englishman had to stop firing because we were in the most helpless situation, namely fast diving, between the English ship's boats.

Our two attempts to torpedo were unsuccessful because the English had now been warned and were able to outmaneuver the torpedo.

"Unfortunately, despite all efforts, the submarine was not sunk," says my opponent at the time, the commander of the Q-ship "Margit", an English submarine trap, Lieutenant-Commander G.L. Hodson in his report "Fortunately we weren't sunk," I was able to report. This incident had one good thing for us; it was a salutary lesson that the enemy had given us. Equipped with an infallible “nose,” we went about our business from then on and sank steamer after steamer. It was my first and most valuable war experience. This could not be achieved through theory, only through practice.

One of the toughest men I clashed with was the commander of the English submarine hunter "Primula".[49] The "Primula" was a completely new type of ship that appeared in the Mediterranean intended to eradicate the "submarine plague". Looked like a cruiser, was well-armed, had depth charges, and had little draft so that she would be shot under, and also had good underwater layout. She could take a lot of torpedoes. But we didn't know any of this when we got tied up with her. My first torpedo hit her in the forecastle, causing the foremast to tip forward. Since she could no longer proceed forward, she drove backwards at high speed in order to ram us, so that we had to disappear as quickly as possible. She would shoot like crazy when I showed my periscope - and I had to, because I couldn't give up the fight so quickly! I attacked and fired a second time, but didn't hit because my opponent cleverly outmaneuvered the torpedo. The same thing happened to me with the third torpedo. But I was so determined that I would never have given up. With a heavy heart I sacrificed a fourth torpedo, which finally (the fight had lasted four hours) hit the "Primula" in the stern and made her unable to maneuver. Now she had had enough, I could see that, so I could drive away with peace of mind. From the newspapers we saw that after another three hours the ship had sunk and its crew had been rescued.

The breakthrough through the Strait of Gibraltar was not always particularly pleasant, but it had to be attempted in order to catch Atlantic trade. There were powerful searchlights on the Rock of Gibraltar that were constantly scanning the narrow area for submarines, and there were also submarine hunters and destroyers that were supposed to hunt us down and force us to dive. If you initially attempted underwater, then with your low speed you

---

[49] Lieutenant-Commander Claude Baynard Evans

would not be able to cope with the strong current that was constantly moving from outside to inside; Besides, no one knew what kind of mess the enemy had prepared underwater. That's why I always traveled surfaced at night and put up with the disgustingly bright searchlights. Unless there was a destroyer nearby, it wasn't that bad. It was best to go hard close under the African coast: for there the current wasn't so strong and you had a good view of the route against the bright searchlights.

So I once again decided to advance towards Atlantic because the western Mediterranean had been swept clean. I took five English steamer captains with me as prisoners; they would be handed over latter at Cattaro. Their fate was linked to ours. Their lives depended on how I ran my boat. No wonder we grew close. As we proceeded through the strait, they became restless, they must have known the dangers, perhaps better than I did. – The sun sank behind the protruding rock of Gibraltar when we tucked ourselves under the African coast at Ceuta. Everything was clear on the route, but we couldn't see what was hidden to our left under the looming dark coastal mountains. The watch in front of me on the tower was manned by the best eyes we had: in front of me to the right the senior watch officer, Oberleutnant zur See Loycke, left our second seaman[50], Boatswain's Mate Timm. Behind me are two more sailors. The entire crew is at combat stations ready to carry out any command without hesitation. The engines were at their best and made a hell of a noise in the quiet night.

"Destroyer two points to port, very close by!" shouts Timm. And then the black shadow comes rushing in, it looks as if it wants to ram us, and then it passes about 50 meters in front of our bow. Like a huge ghost. So close that we can

[50] Seemännische Nr. 2 in the original. The second most senior NCO onboard.

hear his ventilation machinery and even English command words. We are at a loss for spit. The main danger was now over, so I stayed up and didn't dive. Apparently the destroyer had been lying in wait under the coast, perhaps saw us briefly in the searchlights, certainly heard our engines and then missed us in the ramming attack. Now he was clearly silhouetted against the strait so we could see him clearly as he raced around searching. When I turned around, I saw a second vessel on the port side against the dark coast, which apparently signaled our location to the destroyer, but was probably at anchor because it didn't come and was content with illuminating us with its searchlight, which we easily dodge. When morning dawned, the Straits of Gibraltar lay behind us and the Atlantic Ocean took us in, where we made good kills.

It was a Sunday afternoon. We were on our return voyage and were south of Sardinia in fine weather. We were just about to sit down in the forward compartment for a Sunday coffee, which our clever "Schmut"[51] Bölts had brewed from good beans we had captured, when a large steamer with three funnels was reported to me. It came up very quickly - so it was running at high speed. He also ran zigzag courses, but had no escort with him. I followed his movements while submerged, but only had a chance of hitting him if he passed near me. The attack was difficult because we only had one torpedo left, and it was in the stern tube of all places. We had to be very lucky if we were able to strike it. The ship was quickly approaching; it was painted gray, a passenger steamer with large promenade decks. And we weren't allowed to shoot it down as there could perhaps be an American on board. In any event, I carried out the attack; I wanted to make the final decision

[51] Dirt in German. Affectionate slang for cooks in German torpedoboats and submarines where the work environment coexisted with considerable filth.

whether to shoot or not depending on my final impression. I already believed that the shot would be of no use, since I couldn't get into firing range with his zigzag course. Then suddenly he turned towards us - the torpedo could not be deployed in the normal way, only as an angle shot, that is, Then suddenly it turned towards us - but the torpedo could not be deployed in the normal way, only as an angle shot, i.e. after being launched it had to turn according to the previously set angle and scoot towards its target.

Everything happened very quickly now. The final impression was that it was a troop transport, although no troops could be seen on deck. It was instinct again. He passed at a distance of 900 meters. Aiming was difficult. “Torpedo Go!” The boat vibrates when it is launched. I watch. The torpedo takes its course according to the presets and runs towards the steamer.[52] At the bottom someone is counting the seconds. You can hear a pin drop in the boat. It is disappointing when the man keeps counting and nothing happens. But after 40 seconds there was a sharp metallic blow - the warhead pistol - then the crack of the detonation. A resounding “Hurrah!” sounds from below. Through the periscope I see a tall explosive cloud rising, first 10 then 50 meters behind the stern, so that for a moment I believe that the enemy has thrown a depth bomb. That's how fast the ship was going. It only gradually loses its speed and very slowly begins to sink deeper aft. I still can't believe one torpedo is enough for this huge ship.

Now there is a terrible mess on the deck of the ship. Crowds of people are crowding onto the upper deck, some are trying to clear the guns as a final act of duty, and many leap into the numerous boats. These were thrown down so hastily that they got into the water the wrong way and

[52] The preset was 90° and the torpedo was set to 3meter depth.

swamped. People hang on the boats. Everyone wants to get on board first. It was horrible to see. I had had enough and had my men come into the tower one by one and take a look through the periscope. Some stare, motionless, others' faces are filled with horror. They came upstairs quick and curious, thoughtfully and seriously, everyone took their place again.

The mighty ship sank deeper and deeper until it stood out against the glowing red evening sky, raised its shiny bow vertically towards the sky, stood there for perhaps 1-2 minutes, and then shot straight into the depths. A magnificent and horrifying image that I couldn't get rid of for a long time. 20 minutes after the torpedo shot there was nothing left but a field of rubble with a number of overcrowded lifeboats that I could do not rescue. We sailed away into 30 meters of water, each one lost in subtle thoughts.

We heard more details when we arrived in the next few days. It was the "Gallia", the largest French troop transport, carrying senior staff and troops for the Salonika front. 1,852 men went down with him. “France's worst shipping disaster since the start of the war,” wrote the French newspapers. France had lost a battle. Only now did the joy of our success take over. — Chief Helmsman Neumann and his watch were particularly proud, as they had seen the ship first and were therefore able to claim success for themselves. Because there was lively competition between the three watches, who diligently kept records of the ships whose sighting led to them being sunk.

One day, while my boat was still being repaired, I was called to Berlin and given an extraordinary assignment: U35 was to go to the western Mediterranean and arrive in Cartagena with an imperial hand written letter to the King of Spain in thanks for the good treatment of the Cameroon combatants interned in Fernando Po and with 35 boxes of quinine and other high-quality medicines for them. It was not possible to provide aid to the Cameroon men in any other way. It was also intended to test how the Allies would behave towards a neutral country if it were to grant the normal protection under international law to a submarine that, according to world opinion at the time, was excluded by all international law. Of course, I should under no circumstances exceed the internationally prescribed length of stay of 24 hours. I was advised to be extremely careful when going in and out. In the harbor I should be prepared for acts of sabotage by enemy spies.

There was some pretty good loot on the way there. One day before the scheduled arrival date we ceased waging war and brought the boat into a state of peace as best we could. It was a strange feeling when we gave up the harsh craft of war for a day and set out on a pure peace mission in the middle of war. — The Spanish harbor pilot was somewhat surprised when U35 unexpectedly appeared between the pier heads of the harbor near Hellwerdcn on June 21, 1916 and turned into the harbor of Cartagena in order to dock alongside the German steamer "Roma" that was moored there. There were no enemy warships in the harbor. This relieved me of one worry. — When our national salute of 21 shots rolled across the harbor at 8 a.m. - the first and only one ever fired by a submarine - the medicine boxes had already been discreetly handed over to the "Roma" and our naval attaché from Madrid was on his way to Cartagena - to receive the handwritten letter. Official visits to the local dignitaries were quickly completed. Our German consul Dr. Tell accompanied me.

The Spanish people were full of warm friendship for Germany. Crowds lined the harbor until late at night to see U35. With the permission of the authorities, I then laid alongside the Spanish cruiser "Cataluña" where I was safe from acts of sabotage. Given the rich hospitality that was offered to us, I had to allow the few Germans, but also the many Spanish officers, to inspect the boat, whereby strict control and cordoning off of the approach of unknown boats were carried out. The crew was showered with gifts. At 2 a.m., after a 22-hour stay, U35 cast off and left the harbor to roaring cheers from Spanish sailors and demonstrations of sympathy from the crowd who were still holding out. – Since you could see the glow of many lights outside, an exciting fight was now expected. Outside the pier heads, I turned off the lights and went underwater to give us some rest after the hustle and bustle of this tiring day. And since our enemies had made every effort to intercept us in front of the harbor, we only managed to surface again in the late afternoon to begin the return trip.

In addition to the grim seriousness of the war, we also experienced some humorous changes. This was very much welcome to improve our mood. Our on-board monkey Fips, a member of the vervet monkey genus, was simply priceless for this purpose. We picked him up at the last moment from the deserted dinner table of a sinking steamer whose crew had long since fled. "Obermüller, go on board again and save the monkey," I called to the officer on watch. But the monkey didn't want to leave his delicious meal. He didn't seem to enjoy climbing down the vertical side of the Jakobs ladder at all. And so he kept biting the hand of the poor Obermüller who was dragging him behind him until, when he got down in the dinghy, he dunked him hard into the blue sea, which had a visibly calming effect on Fips.

Fips quickly became the favorite on board because of his funny pranks and affection. He was given quarters in the stern torpedo room, where he was very close to Böltg the cook, from whose hands he stole many eggs intended for the pan. — Since Fips was on board, there was always something going on. He also liked to grace my berth with his visits, eating my pencils, tearing up my radio messages and occasionally drinking the empty inkwell. He often accompanied the watch on deck, carried in the side pocket of Qbermüller's leather watch jacket, from which his little head poked out cutely. Malicious tongues claimed this chubby bag must have been wet from time to time - - Fips entertained us with his ever-new pranks for almost a whole year until Haiungs later donated him to the Berlin Zoo.

The Bocche di Cattaro, that picturesque bay on the Dalmatian coast at the foot of Lovcen was our base, where, under the protection of the Austro-Hungarian coastal fortification and mine barriers, we overhauled our boats, filled them up to the brim again with oil and ammunition and fed ourselves some. We also desperately needed this. There was an accommodation ship where you could sleep undressed again. You could indulge in water and soap to your heart's content. And there was also variety in the food. And finally you could really run and climb around in the mountains.

During the 4-5 weeks that we were out at sea, we lived in dirt and grease, man, officer, commander. We were all the same in our gray leather package. All covered with a sticky patina consisting of sea water, oil and dirt, because water for washing was scarce. There was a bucket in the officers' compartment in which the officers could wash their hands. Each week everyone was only given a small sink full of delicious water.

And as far as food was concerned, we lived off the enemy as long as he still had some, not badly, sometimes even very well. But when things began to become scarce in 1917, we resorted to the iron stocks our homeland sent us. There was still some left over for us submariners, but after the brown bread had been in the well soldered tins for several weeks it was already slightly moldy. Then the hard bread was brought out. It was so hard that even the mold couldn't harm it. To do this, you had to knock out the worms before you could "process it further", that is, with the help of a sledgehammer, meat grinder and coffee grinder, turn it into fine flour, from which our skilled "Schmut" (cook) baked the most wonderful rolls Baking imaginable.

It would be ungrateful to ignore the fact that we also had good days, which were extended as long as possible thanks to economical storage. When the torpedoes were fired and the torpedo tubes were empty, they were stuffed with sacks of beautiful, white wheat flour, which the enemy no longer had any use for, but which we in Germany only knew about as a rumor. 2-3 hundredweights of flour could easily be stored in each tube. And so on special holidays there was beautiful, crispy crumb cake or occasionally Sandtorte with coffee.

But now the memory of these culinary delights has completely distracted me from my military activities, which I have to report on.

Before leaving Cattaro, final tests were carried out, engines were run in, the boat and crew were trimmed, diving and depth control tests were carried out and the boat was then clearly reported for departure. There were still the farewell letters that each of us wrote to our relatives. It always said the same thing: They were going to sea, they would be back in 4-5 weeks, they shouldn't worry, nothing could happen, etc. It was then written down like that. But, hand on heart, my comrades! Each of us felt a little queasy on the day of departure, everyone was busy with their thoughts. There was a great seriousness in the faces of all of these men.

Deep down inside sat the "human bastard"[53] that had to be overcome, there was the anxious worry of whether we would be able to get back or whether fate would take us too like so many comrades.

In such moments you felt the weight of responsibility for the well-being and woe of the 35 brave men, for the weapon that was given to you to fight for Germany's freedom. But nothing was shown to the outside world. But once the first obstacle, the Otranto Strait, had been overcome and the first enemy ship had been sunk, then the crew felt a victorious mood, as if no one could harm us. And it was fine like that. Because you can only achieve victories with optimists. And then if fate gave you a blow and you got hit in the nose, all you had to do was eradicate any pessimistic mood that might arise with a new successful attack.

We had to go through Otranto Street twice, on the way through it and on the way back. The enemy knew that. Wasn't it obvious that the entire French and Italian naval power and part of the English one with their numerous forces and resources should deploy here to seal off the Adriatic to the south! And doesn't it seem simply incomprehensible today that for three years the few German submarines were able to inflict such immense damage on Allied trade in the Mediterranean (over 3 million tons were sunk) without the combined Allied forces managing to significantly impede passage through the Otranto Strait until the end of the war! (4 submarine losses in the southern Adriatic.)

Both sides of the strait were flanked by excellent bases. The sea was often littered with guards, dragging their nets

---

53 Menschliche Schweinehund in the original. Reference to German allegorical expression "Innerer Schweinehund"-the inner weakness would obstructs the individual from carrying out an unpleasant task.

riddled with mines. If you got among these, they would literally ensnare the boat, the mines would cause the boat to leak, the nets would make the screws unclear. If you got a leak, you would sink. And 900 meters of water depth, that was no fun! If you came up with tangled screws, the enemy would give you a warm welcome. Then it was over. There were destroyers and almost invisible submarine hunters swarming above, against which you couldn't do anything, because you were incapable of maneuvering. Then the only thing left was: Surface! Abandon Ship! Sink the boat! Whether one received medical treatment or was rescued as a prisoner of war depended only on whether or not the humane aspect of warfare was still being observed by the respective enemy because of the monstrous incitement stirred by the Allies. During the day, aircraft provided surveillance.

Each of us commanders had his own method of overcoming the Otranto Strait obstacle. Some clung to the Italian coast, some to the Albanian coast near Saseno, others closed their eyes and went to 40 to 60 meters and sailed at that depth through and still others chose night and wound their way above water between the destroyers and guards. I was one of the latter. I preferred avoiding the unknown dangers lurking underwater and breaking through in the dark night with a full view of the water. We often darted past within a few hundred meters of the guards or drove alongside the unaware destroyers just to avoid having to dive.

Another, much more serious danger were the enemy submarines that lay in wait for us in the southern Adriatic between Cattaro and the Otranto Strait, and with which this sea area was literally paved. Of course, these colleagues of ours had to know best how to deal with us. But they really weren't to be envied. Having to lie in wait day after day, going underwater for around 18 hours in the summer. Only up at night to charge the battery and get some fresh air. And how many times must they have lain around without ever seeing a target? When one actually came into view, it wasn't so easy to approach and hit it. The main difficulty for them was not to shoot each other. Because a U-boat is a submarine, and regardless of whether they were friends or foes, they looked incredibly similar. No! The German submarine arm was better off. We found targets everywhere. Attacking them was often dangerous, but certainly never boring!

And so of course the enemy positioned his submarines where we were sure to pass. Especially immediately before the entrance to Cattaro, but also on the Cattaro-Otranto Straits route. This was not possible in the Otranto guard area. Because every submarine that showed up there was attacked. We became more cautious when we got to know these new methods and our comrades acknowledged them. On the one hand you only went in and out when it was dark and on the other hand you stayed mainly under water in the endangered area during the day, i.e. You should actually do the latter. But it wasn't always done. Because if you had passed the dangers of the Otranto Strait at night on the way back, you didn't want to be underwater the whole next day, but were in a hurry to get to the harbor that same day. This occasional imprudence almost cost us our lives, as we will see later.

We inadvertently clashed with our enemy brothers three times.

Once, in February 1917, we had just left Cattaro in the dark in the evening. The two Austro-Hungarian torpedo boats that had escorted us out through the barriers disappeared into the darkness after signaling us with a “Happy voyage and good luck.” We drove south. The engines went “Ahead

Full." On the tower watch officer Oberleutnant zur See Obermüller with his two men and me. The off watch was already lying in the roses and sleeping. The weather was good, everything was peaceful. Surprises were hardly to be expected, at least not in this area.

Suddenly Obermüller shouts: "Hard to port, both engines Ahead Emergency Power! A submarine ahead!"

Glass to the eye. Damn another one! Very close to our bow in the ramming position, a gray shadow slowly passes by. An enemy submarine that we're just getting clear of and that was stuck around here at low speed. It was not possible to use weapons; because as quickly as the brother came into view, he disappeared again. If we had paid less attention we would have bumped into him; And since we had loaded our torpedo tubes forward with live torpedoes, they would have detonated and both boats would almost certainly have been blown up.

A month later we departed by day and were just free of the minefield. This time Chief Helmsman Neumann was on watch. People sat eating. We surfaced and drove at high speed because we wanted to go through the Otranto Strait that same night." We were very careful. Because attention was everything. Suddenly Neumann commands:

"Torpedo track to portside! Hard to starboard!"

That's right, there was the discharge swell and a bright stream of water, the torpedo track, was coming towards us. Thank God, a bit far, about 1500 meters. But good intersecting angle at 90 degrees. A fine shot at least. You have to give that to him. Our eyes were glued to these bubbles that would decide our fate. Does he hit or doesn't he hit? You have a strange feeling in those moments when you are completely powerless and at the mercy of fate. But our boat had turned well. We are clear again. The torpedo went by 100 meters behind our stern. Although we immediately fired our cannon in the direction of discharge swell, we were clear that this no longer made any sense and that our colleague would simply smile at this pointless display of force. Just in case, we waved goodbye to him. As we know today, this was the French Lieutenant de Cambourg, commander of the submarine “Circe”.[54]

With a volley of torpedoes he might have caught us, and it would have been worth it for him. This undertaking cost the Allies at least 24 ships with around 80,000 tons.

Another time we returned from an operation which lasted five weeks. We had made the western Mediterranean unsafe, advanced through the Strait of Gibraltar into the Atlantic Ocean as far as Madeira and had fired all torpedoes. The enemy had been able to follow us closely on the way back through radio messages from his defense network and we had to be prepared for a warm reception at Otranto Strait. We encountered destroyers as we approached them but we remained hidden from them.

Around midnight on November 5th and 6th, 1917, we found ourselves in the narrows, surfaced of course. The engines were going fast. The crew at diving stations. At the top of the tower the watch is equipped with binoculars. The weather is good, we would have preferred worse. And the guard vessels were there with their nets. First one, then two and more and more, across the strait. That could be cheerful. It didn't seem like a good idea to push through, and certainly not to dive. I wanted to avoid them. We proceed towards the east. They didn't stop there. We counted 15 units. So turned west towards the Italian coast. Thank God! There was obviously a gap in the barrier. Now

---

[54] French sources indicate Circe’s attack occurred on March 31 not February.

to get through it! — And we succeeded. — Now some star signals went off by the guards. They probably saw us after all. Just hold on now, just don't dive! And after we had avoided a few more dimmed vessels, it slowly became light. Nothing was in sight anymore. Far away on the port side is a strip of the Italian coast, on the starboard the high Albanian mountains with the offshore island of Saseno, a first-rate Italian base.

The sun went out and a wonderful morning dawned. A clear sky above us. A light northwest rippled the blue waters of the Adriatic that spread out in front of us. A slight swell running from the north occasionally let the bow of U35 into the sea behind.

I crouched in the tower and, bleary-eyed as I was, I shivered. But we were in good spirits. Once again we had done it and brought home a really nice bounty. 11 ships with 30,000 tons. You couldn't ask for more in these times of tightened defense. And while the new watch went up at 8 a.m. and the one which replaced them rumbled happily down through the tower hatch, Bölts, the cook, brought me a nice coffee, such that a pleasant warmth spread through my limbs.

We all had a strong longing to come home. 35 days at sea in dirt and grease is no small feat. 35 days wearing the same clothes!

35 days of only sleeping for hours with heightened senses, ready to rush up the tower at any moment. You almost had the feeling when something was going on upstairs. And how did we all look in our oily leather gear, our faces covered in a crust of dirt, unshaven of course! Finally the harbor was within reach. The prospect of a warm bath, a set table and also a lot of alcohol. Yes, you had to have that and then sleep, sleep, sleep again without constantly being

woken up! And what happened in those five weeks? How much mail would you find? And then our comrades! Who would have bought it now? Many a friend had already gone. One would never know where. These were the thoughts that occupied each of us.

But I still had to expect enemy submarines. We had already been shot at, but not hit yet.

I thought: If we continue surfaced, we'll be in Cattaro harbor when it gets dark. Then it's done. We could only avoid the submarine danger if we dived and stayed underwater until evening. Then we would arrive the next day at best. That meant another night outside. The temptation was great, the decision difficult.

If I dived now, my crew would certainly be very disappointed. And finally, you can never wage war with 100 percent certainty.

So I decided, "Stay surfaced! Going in with everything the engines can produce!"

The new watch was up. At the front of the tower stood the watch officer, Oberleutnant zur See de Terra, plus Oberleutnant zur See Prince Sigismund of Prussia, the son of Prince Heinrich. He used his father's influence on the submarine arm in order to receive frontline training. And on both sides aft there were two more sailors on watch. I think they were Oberstadt and Werschkull, both compatriots from East Prussia. Standing at the loaded cannon, ready to fire, was the Seaman Nbr.2., Boatswain's Mate Berger. He was to immediately open fire when a periscope appeared and thereby prevent the enemy from firing torpedoes.

“Now, gentlemen, be careful,” I tell them, “the safety of the boat depends on your eyes!” “And you, de Terra, be ready to turn and open fire. – Under no circumstances dive!” –

“Should I run zigzag courses?” he asks.

“Oh no, the main thing is to be careful. The zigzags don't save us either. – Well, I'm going down below, I'm going to lie down some more. If anything is spotted, please report it immediately!”

Of course they were happy that the "Old Man" disappeared. I climbed down through the tower hatch. There was already a lot of activity there. The last bit of fresh water had been divided up, and the men were standing in every corner to scrape off their sea beards and take off their oily-smelling blue gear. Because we strictly stuck to it on board, no matter how dirty we were in the outside. But when we came in, the two masts full of pennants, each signifying a sunken ship, the crew stood proudly on deck, dressed to the nines, as if they were on parade. And no one could tell that we hadn't gotten out of the dirt for five weeks.

“When are we arriving, Herr Kaleu,” Kohrs, the chief engineer, asks me.

“Tonight after dark. Let your sewing machines run properly. Then maybe it will be sooner."

A look into the individual compartments. I see beaming, smiling faces everywhere. Then I hit my bunk and try to fall asleep, but that's out of the question, even though the men are trying their hardest to scurry past my bunk as quietly as possible. Nerves are just too over stimulated despite the tiredness. I wanted to be fresh again after the night awake because in the afternoon outside of harbor is where the submarine danger was assessed at its greatest.

I had been lying there for ten minutes when I was startled by a deafening bang above me on the deck. I literally collapsed. — The devil knows what they had done up there! — I was on the tower in just a few steps.

"What's going on here?"

There stood the watch and my capable Chief Helmsman Neumann, a senior experienced submariner, who was about to make his navigational observations, every sign of terror on their faces. Speechless, their eyes wide, they stared transfixed at a spot on the side of the ship's side. I follow their gaze and see, just a meter away from us, the glass lens of a periscope, which stands steadfast out of the water, staring eerily at me and then being turned slightly as if by an invisible hand. I can almost grab it with my hands.

Our boat was in full rotation, so our stern came within a few centimeters of his periscope. Instinctively, de Terra had thrown the boat around, although there was little point in the maneuver since no force in the world could save us from the oncoming torpedo. The next moment there was bound to be a crazy racket and all of us, dismantled into our component parts, had to whirl around somewhere in the pleasure.

Then the torpedo path passed the boat exactly in amidships under the tower - and continued out the other side.

Nothing happened - Undershot - Maybe a few centimeters below its target. Another torpedo followed on the heels of the previous one, so to speak, and went under the stern. We were able to easily avoid another one that came up from aft. The periscope, however, remained missing.

This all happened in just a few seconds.

And now de Terra reported, his face still white as chalk after he had found his voice again:

Suddenly, about 50 meters from the helm side, the sea turned white-green - a discharge swell! Behind it the periscope. Two torpedo tracks were coming towards us at breakneck speed. He was still able to give the commando "Hard to Port! Emergency Speed ahead!" Then, two meters from the ship's side, a giant steel eel with a red head - a torpedo - shot silver-white out of the water, as if it wanted to go towards the tower. Its propellers, which were now in the air, howled loudly. As if in defense, they instinctively raised their right arm. Then the torpedo hit the deck with a loud bang and slid along the steel plates with a loud clink, right between the tower and the gun. A splash in the water on the port side and then he kept running as if it had to be that way. Then I appeared at the top and saw the rest myself.

I must have looked at my officers with some disbelief, their statements seemed so incomprehensible. Literally they showed me the railing on the gun that was bent downwards, as well as dead traces of paint on it and on the tower's breakwater.

It was lucky that Berger wasn't standing at the gun. That would have been a wet death ride. And it was a miracle the torpedo flask hadn't exploded due to the deflection. If the firing pin attached to his head had touched even part of the boat, it would have been over.

All went well once again and we were soon backing on course to Cattaro.

I disappeared below and couldn't help but remark to de Terra that, as my torpedo officer, he could at least have determined the type and number of the enemy torpedo.

Meanwhile, those down in the boat also realized something extraordinary must have been going on above. When they were told that a “friendly” colleague had fired a barrage of torpedoes at us and the eels had moved not only below us, behind us but also above us, they thought those above were slightly crazy. In disbelief, one after the other did gymnastics on the deck and felt the bent railing. Only then were they convinced.

It's only natural that no one wanted to believe us about this experience. When I reported the boat back to my flotilla commander in the evening, he must have thought I was completely crazy too. Only the railing gave him back his belief in my normal state of mind.

Our boat became a place of pilgrimage those days. They came in droves from all U-boats and Austro-Hungarian ships to look at the railing. I was also able to welcome Hungary's current Imperial Administrator Horthy, then commander of the "Novara", a loyal friend and patron of us U-boatmen, on board.

It was clear to each of us: The good Lord had once again been at the helm, or as others said, “held his broad thumb between them.”

From now on it was a foregone conclusion on board: we were just lucky. Nothing could happen to us now. It was presumptuous to say this, but it was true.

At the time, I felt deeply sorry for my enemy colleague because of his bad luck! How long the poor man may have been lying in wait. And then the longed-for enemy came in front of him and - - In my excitement I forgot to give him a military salute when he came face to face with me at a distance of one meter with his periscope.

Long after the war when I was with the "Emden" lying in Reapel and talked about this incident, my Italian comrades remembered it and told me that it was the French submarine "Faradon[55]", whose commander, Lieutenant de Vaisseau Bougard is said to have asked to be replaced on the grounds that a submarine commander who was as unlucky as he was should no longer allowed to lead his boat. But that could also have been a fairy tale.[56]

In his war diary there are the following entries about this incident:

> November 6, 1917. 8h35 Lance 3 torpilles sang resultat par 41°18' et- 18°45' a moins de 100 meters. (8:35 a.m. 3 torpedoes fired without success in 41 degrees 18' north latitude and 18 degrees 45' east longitude at a distance of less than 100 meters. signed Bougard.)

Apparently we had counted one torpedo too many. But three are ultimately enough.

For two years, from the end of January 1916 to March 1918, I fought in the Mediterranean with U35. Officers, deck and non-commissioned officers changed significantly because more and more submarines were being built. The rest of the crew essentially stayed together. We only thought in terms of tons. We had conversant with Lloyds Register (list of all merchant fleets). We would have preferred warships. But they sailed sparingly. And it was particularly lucky if you came across one. However, the destruction of tonnage used by the enemy was infinitely more important for the success of the war. That was clear to all of us. "Sweep them

[55] Typo in original. Faraday correct spelling.

[56] Henry Dominique Bugard commanded the submarine "Faraday" from March 13, 1916 to April 25, 1918. Von Arnauld does not list a third attack and the dual attempt by Faraday may be considered as three in total.

off the Sea!" was the catchphrase. We searched every corner of the Mediterranean restlessly. Our voyages continued into the Atlantic. Our fighting methods were in every way similar to those of the enemy.

Once we sank 56 ships weighing 90,000 tons in a four-week voyage. We didn't have a torpedo, a shell or an explosive cartridge left on board. No wonder we were soon in first place in terms of the number of sinkings. There were awards to be given, dares to be awarded. At the beginning of 1918, the new submarine cruisers were ready to take the war to the American coast. A new operation there was appealing. In May 1918, our Kaiser appointed me in a personal handwritten document as commander of U-Kreuzer U139, which was to bear the name of my fallen friend and comrade "Kapitänleutnant Schwieger". And so I completed my work on U35 in March 1918 with a total of 200 ships of 500,000 tons, most of which were named. But the worst was yet to come: that meant being separated from my good boat and some of my men. Some had to stay behind to keep U35 operational. I was allowed to take half of my crew with me. I will never forget that farewell on March 17, 1918. Not many words were exchanged, but some of tough sailors' eyes watered.

U139 was no longer a boat, but a real underwater ship of more than 2000 tons of water displacement with 8 officers, 100 crew, two 15-centimeter guns, 1000 shells and 20 torpedoes. She was able to stay at sea for four months without having to replenish fuel and supplies. After a stormy three-week journey, we found ourselves in the Bay of Biscay on October 1, 1918, to experience an adventure the likes of which we had never experienced before. We caught a convoy of 10 steamers, which was protected by two auxiliary cruisers and a swarm of small guards. Since my heavy ship wasn't maneuvering underwater as quickly as the convoy was zigzagging, I suddenly found myself in the middle of the convoy and my torpedo went past.

This was unsuccessful. But after all, why did I have my big guns? So I wanted to try again above water. The convoy passed over me. When he was 5,000 meters away, we appeared and for the first time our heavy shells rushed towards the enemy. But now hellfire broke out on us too. The water columns rose up around us. Thank God, without hitting. But now the leader cruiser[57] was roaring up, so I had to let go of the steamers and concentrate on him. I caught the right moment for the diving maneuver. As soon as the water had pooled over the tower, a bang. The last volley of the auxiliary cruiser hit the tower. Shrapnel hit the tower with force. That went well again. We laughed at the few weapon water bombs that followed. The main thing was that he let go of us and followed his disappearing convoy.

My decision was clear. Give in no? It's out of the question. So attack for the third time but this time well prepared.

It was lunchtime. First something in the crew's stomach. Meanwhile the escort was at the horizon. Then up and after him at emergency power. The convoy sees us coming, it steams as fast as it can and widely spreads out. We have everything ready to fire; there are 40 men alone in battle stations on the upper deck, all of which have to come down if the enemy forces us to dive. On my order, "Permission to fire," the artillery officer, Kapitänleutnant Pistor, attacked the closest steamer ahead. After a few shots, he is hit and turns, apparently unable to maneuver, and launches boats, a sign that he is giving up. "Change target!" to the next one.

---

[57] The Armed Boarding Steamer HMS Perth was escorting convoy H.G.109. Commander Gerald Parnell in command who was commended for his actions. His vessel was with three 4.7" guns vs. the U139's two 5.9".

It doesn't take long before he falls aft. Things would have continued like this if the auxiliary cruiser had not once again involved me in a firefight, in which it received a few hits, but in the long run I lost out. So I had to disappear again. He initially renounced further persecution. He probably had to go to support the two damaged ships. I followed, watching the events through the periscope. Because one thing was certain for me: under no circumstances were the two brothers allowed to be brought in. But one of them sank as I approached. The other was lying on the water with a slight list. Help came from the French coast. In any case, a swarm of submarine hunters were buzzing around him like hornets to make an attack impossible for me.[58]

We had been working since ten o'clock. Now it was starting to get dark. If I didn't succeed in getting a shot at the

---

[58] Extracts from HMS Perth's log follows:
Gibraltar to Milford Haven
Lat 45.2, Long -11.4
4.0am: Ship zig-zagging ahead of convoy
11.45am: Submarine reported, convoy zig-zagging
1.20pm: Engaged submarine with gunfire, range 8800 - 4800 yards. Sub Lieutenant Stevenson RNR killed
1.30pm: Submarine dived
2.40pm: Rejoined convoy
4.0pm: Engaged submarine at extreme range
4.10pm: Ceased fire, submarine out of range
5.20pm: Engaged submarine, range 8000 - 5800 yards
5.30pm: Sub. disappeared, Assistant Paymaster C Maile, RNR, killed
6.30pm: Proceeded to two steamers in distress
6,.5pm: SS BYLANDS sunk
7.10pm: Picked up crew
7.10pm: Lowered whaler to examine SS MANIN, abandoned by crew
7.15pm: HMS P67 & HMS P69 joined
7 20pm Set course N46E speed 80 revs, proceeding towards convoy
7.30pm: Set course towards convoy. Port fore bridge deck awning destroyed in action, remnants used for burial of Sub Lieut Stevenson & of Paymaster Maile
9.00pm: Stopped to put patch over shell hole on port side forward

distressed vessel, it would certainly be towed in during the night. And so I made 4-5 attempts. Whenever I was within shooting distance, a beast of a submarine hunter would rush overhead and I would have to dodge downwards. By now it had become completely dark. — Off watch personnel were sitting at dinner. The only men involved were the engine crew, torpedo crew, depth control staff and in the tower with me, my capable Senior Helmsman, Leading Seaman Otto as combat helmsman, and Senior Radio-Telegraphist Plate at the engine telegraph post.

Now came the final attempt. Screw noises approach. Go deeper. Up again. In front of me is the black hull of the steamer. "First bow tube Go!" And the torpedo roars off. I accelerate to the steer-to depth and say down to the Central: "Somewhat close. Hopefully it doesn't fall on us." Then there was the well-known detonation and a hurrah from below that was drowned out by the terrible crash and noise above me. The periscopes shake. The light goes out. At the same time, the boat leans heavily towards port, masses of water flood into the tower above us, and it only slowly stands upright again. And in between you can hear the crash of depth charges all around us.

My first thought is the ship we just torpedoed must have fallen on top of us while it was sinking. Since such a rapid sinking is actually rare, the steamer must have already sucked in a lot of water through shell holes.

The light comes on again. The pumps are turned on. The pointer of the depth pressure gauge moves deeper and deeper. The torrent of water that falls on us through the leaks in the tower overhead and through the cracked tower hatch becomes ever stronger. The water level in the periscope shaft slowly rises higher. "The pumps can no longer handle the water," reports the "Leader." Marine-Oberingenieur Fechter, from Central. So we sink. The sinking steamer takes us into the depths. Not a pleasant view at 1000 meters of water! My hair stood on end at the thought. The only other option left is our last resort: use compressed air to sell our lives as dearly as possible. If this happens, some of my men will be imprisoned. I still have a small glimmer of hope. “Senior Engineer in the tower.” — “We will bring the ship to the surface with compressed air. If this succeeds, we will dive again immediately, but only go deep enough so that there are only a few centimeters of water above the tower overhead. And then I will to try to get out of the mess." The commands follow: "Compressed air on all tanks!" - "Both engine, Full speed ahead!" - "Surface!"

My eyes are glued to the depth gauge. Will the ship obey? Will these last resorts work? A few anxious seconds follow, now the depth gauge pointer is at 40 meters, then a terrible jerking, tearing, grinding, scraping starts above our heads, as if something is slipping from our tower. Our ship leans over to starboard for a moment. And suddenly we rise like a balloon. The inrush of water is slowly subsiding. We could breathe a sigh of relief even though depth charges were still crashing around us. We broke through the surface and quickly went back down, forestalled our ship surfacing by opening the quick vents. —- Go to 10 meters! — Finally the pumps were able to hold the incoming water. The periscopes were in disarray and unusable. We were blind. We crept away just below the surface of the water, expecting at any moment to be run over by one of the rushing guard vessels. Minute by minute passed slowly. The detonations of the depth charges became weaker and weaker and finally stopped completely. After an hour we surfaced. Deep night surrounded us, thank God! Nothing could be seen, only in the south, where we came from, a few searchlights flickered through the night.

Now we could take a closer look at our boat. The tower superstructure was shattered and the dent in the steamer's bow was clearly visible. The upper deck is dented. The turret hatch was bent and couldn't be opened, and worst of all: all three periscopes were bent over and hanging down like broken asparagus. The entire upper deck was littered with shrapnel and debris. – The anxious question before me was: How was I supposed to bring our U-cruiser with its good men home safely in this condition? Would we ever get out to the enemy again given the current war situation? We all had to sleep now. Then I would see further.

The next day looked friendlier. It brought us a small sailing boat first thing in the morning.[59] We then found a barrel of cement with which the Chief Engineer Fechter and his brave engine crew managed to seal all the cracks and leaks in the tower overhead with concrete. At least we were able to dive again without endangering our lives. However, we remained blind underwater.

Nobody thought about going home anymore. If we could no longer fight underwater, we at least wanted to take our 1,000 shells to the enemy on the surface and send down what could be brought down. Our courage to live was unbroken. If only there hadn't been the bleak radio messages from home that our thoughts revolved around!

We later had a hot battle in the Azores. We chased a large steamer, but it was faster and escaped us. In return, his companion, a small Portuguese gunboat, attacked us, which was far inferior to us in terms of artillery. Nevertheless, it fought with unparalleled bravery, unable to avert its fate. Half of the approximately 40-man crew including the commander were killed.[60] We took in the

---

[59] The Portuguese flagged Rio Cávado.

survivors, bandaged them and ensured their unhindered return home. To the wounded first officer I could only express my respect for their heroic fight.

That was our last fight. On October 21st, the news of the cessation of submarine warfare came via radio telegraph. We had to put the weapon down in silence. We didn't like going home.

The red flags of revolt were flying in Kiel when we arrived on November 14, 1918. That was the bitter end.

---

[60] Primeiro-Tenente José Botelho de Carvalho Araújo in command of the armed trawler NRP Augusto de Castilho in the battle.

Made in the USA
Monee, IL
04 November 2026

67379299R00085